Landscape Interpretations:

History, Techniques, and Design Inspiration

Join us on the web at

agriculture.delmar.com

Landscape Interpretations:

History, Techniques, and Design Inspiration

PAUL C. SICILIANO

THOMSON
DELMAR LEARNING™

Australia Canada Mexico Singapore Spain United Kingdom United States

Landscape Interpretations: History, Techniques and Design Inspiration
Paul C. Siciliano

Vice President, Career Education Strategic Business Unit:
Dawn Gerrain

Director of Editorial:
Sherry Gomoll

Acquisitions Editor:
David Rosenbaum

Developmental Editor:
Gerald O'Malley

Editorial Assistant:
Christina Gifford

Director of Production:
Wendy A. Troeger

Production Manager:
Carolyn Miller

Production Editor:
Kathryn B. Kucharek

Technology Specialist:
John Cacchione

Director of Marketing:
Wendy Mapstone

Marketing Specialist:
Gerard McAvey

Cover Design:
Liz Kodela Design

Printed in Canada
1 2 3 4 5 XXX 08 07 06 05 04

For more information contact Delmar Learning,
5 Maxwell Drive, PO Box 8007, Clifton Park, NY 12065-2919.

Or you can visit our Internet site at
http://www.delmarlearning.com.

Library of Congress Cataloging-in-Publication Data

Siciliano, Paul C.
Landscape interpretations: history, techniques, and design inspiration / Paul C. Siciliano.
p. cm.
ISBN 1-4018-1154-X
1. Landscape architecture—Europe—History. I. Title.
SB470.55.E85S53 2004
712'.094—dc22 2004045288

NOTICE TO THE READER

Publisher does not warrant or guarantee any of the products described herein or perform any independent analysis in connection with any of the product information contained herein. Publisher does not assume, and expressly disclaims, any obligation to obtain and include information other than that provided to it by the manufacturer.

The reader is expressly warned to consider and adopt all safety precautions that might be indicated by the activities herein and to avoid all potential hazards. By following the instructions contained herein, the reader willingly assumes all risks in connection with such instructions.

The Publisher makes no representation or warranties of any kind, including but not limited to, the warranties of fitness for particular purpose or merchantability, nor are any such representations implied with respect to the material set forth herein, and the publisher takes no responsibility with respect to such material. The Publisher shall not be liable for any special, consequential, or exemplary damages resulting, in whole or part, from the readers' use of, or reliance upon, this material.

To my wife, Debbie, for her constant encouragement and unfailing patience,
and to our very special children, Paulie, Maria, and Joseph,
and to the memory of my mother, Anne, my grandmother, Angelina, and my grandfather, Paul.

Contents

PART TWO France

PART THREE England

Preface

My first real understanding of gardens as something more than a composition of trees and shrubs occurred when I was a graduate student at the University of Pennsylvania. I enrolled in a course titled "Reading and Writing the Site/Sight" taught by Professor John Dixon Hunt. This course examined several great historic gardens, focusing on the ideas and influences that led to their creation and the different ways in which people experienced them. I remember struggling at first to understand how gardens could possibly be "read," but soon I realized, after looking more closely, that they do, indeed, have stories to tell. Gardens speak without words. They are an expression of human culture and of one's relationship with the natural world.

When we look back in history we discover that human beings have always altered and controlled their environments, making them more beautiful, orderly, and more expressive of their lives. Throughout history, artists, architects, landscape architects, and gardeners have been commissioned when people have sought to express their ideas, power, and intellect through the arts, especially in the creation of gardens and landscapes.

In order to fully appreciate the value of past work, I am convinced that one first needs to know what to look for. As an educator in landscape architecture and design, I encourage my students to learn from past traditions. I want them to be inspired by achievements of the past, but I recognize that this will only occur when they thoroughly understand what they are seeing. That understanding begins with knowledge of the context around which great gardens were developed. In other words, what were the cultural, political, and economic factors that influenced design programs and how were they expressed? What are the design elements that were used to convey ideas, satisfy functional requirements, or create beauty? What connections can be made between historical background and the design of these gardens?

There are many books available on the subject of garden history. Some are written as surveys, others focus on a particular country or even a specific garden, but few have been written with designers in mind. As both an educator and practitioner in design, I recognize what information is important to a designer's understanding of historic gardens and how that material is best conveyed. I have written this book to facilitate that understanding and to provide the basis for further inquiry to those interested in extending their knowledge beyond the information presented in this text.

Landscape Interpretations examines some of the greatest successes in garden design, focusing on the history that surrounded them, the techniques used to create them, and their relevance to current design practice. It is instructional and at the same time intended as a useful reference, a study of some of the finest gardens in the world presented in a clear and accessible format with plenty of images, associated plans, and time lines. I have chosen to narrow the focus of this investigation to three countries—Italy, France, and England. Each country represents a period of significant development in the early history of garden design.

The gardens of Italy are discussed in Part One of the text, beginning with those created during the great Renaissance period. Part Two examines the royal gardens of the seventeenth-century French baroque period. The evolution of English gardening, spanning several hundreds of years, is covered in Part Three. At the end of each section is a chapter called "Modern Variations" that shows how historic theories and techniques of that particular gardening period have been adapted to a modern design. Essential ideas and theories about the making of the gardens, parks, and piazzas presented in the text are organized around categories including geography and climate, social and cultural history, and landscape expression. Also available is a CD-ROM featuring captioned photos and garden plans, arranged as a virtual tour through the gardens to help the designer further visualize the creation of the setting.

Looking to the past for inspiration is not a new idea; in fact it is very much a part of our professional history. For centuries designers have looked to past achievements, translating successful elements in those designs into solutions for modern practice. The aim of this book is to encourage students and practicing designers to look deeper into the meaning of historic gardens and to consider their potential as a resource and inspiration for current design practice.

About the Author

Paul Siciliano is an Associate Professor of Landscape Architecture at Purdue University in West Lafayette, Indiana. He teaches courses on landscape architecture history and theory, landscape design and construction, as well as an international study course on English Garden History based in Corsham, England. Professor Siciliano also teaches landscape design and history courses at the School of the Chicago Botanic Garden in Glencoe, Illinois. At Purdue his research focuses on the historical development of landscape design theory and practice.

Professor Siciliano holds two bachelor's degrees from Purdue University one in Landscape Architecture and another in Landscape Horticulture, and a master's degree in Landscape Architecture from the Graduate School of Fine Arts at the University of Pennsylvania. Prior to his career in teaching, Professor Siciliano gained over twenty years of combined professional experience in landscape architecture, landscape construction, and wholesale nursery production that he now brings to his teaching program.

Professor Siciliano has written articles on landscape design and history for many trade and popular magazines, including *American Nurseryman* magazine, *Indiana Nursery News,* and *Chicago Home and Garden* magazine. He was a contributor to *Chicago Botanic Garden's Encyclopedia of Gardens: History and Design* (Fitzroy Dearborn Publishers, 2001). He is also a frequent guest speaker at national workshops and conferences organized by universities, botanic gardens, and landscape trade organizations.

Acknowledgments

I could not have attempted this book without the support and assistance of many people. First and most important, I want to express my deepest gratitude to my wife, Debbie, who has always supported me in all that I have attempted. She, along with my children, Paulie, Maria, and Joseph, has spent considerable time apart from me while I researched, wrote, and reviewed this book. A very special thank you to Amanda Chrzanowski, a former student, for her help in the early stages of design and development of the project's CD-ROM. I am enormously grateful to Melissa Hasenour-Miller, also a former student, for her beautiful garden plans featured both in the text and on CD. Her artistry and attention to every detail have made the plans a significant addition to this book. Many thanks to the Thomson Delmar Learning team, who worked very hard to produce a quality textbook. I want to thank the Department of Horticulture and Landscape Architecture at Purdue University for their support, especially Dr. Edward Ashworth, Department Chair, Dr. Michael Dana, and Professor Greg Pierceall, each a friend and colleague. Their advice and encouragement has contributed not only to this project, but also to my professional development. In addition, my thanks to Debrah Huffman and Murray Shugars for their editorial assistance and to Matilda D'Orzo, Simone Carotti, and Paola Veronese for their help in translations. I am also grateful to all of my teachers, past and present, and to all of my students, whose interest in learning is a source of my enthusiasm. I also wish to acknowledge the many garden historians whose research and scholarly publications have provided valuable insight that has contributed to our understanding of garden history. Their work has been a source of my own comprehension and a resource in the writing of this text. And I would like to thank *Chicago Home and Garden Magazine* for their permission to reprint in this book various portions of my articles that have appeared in their magazine.

Thomson Delmar Learning and the author wish to thank the following individuals, who devoted their time and professional experience reviewing this manuscript:

Kenneth Struckmeyer
Washington State University
Pullman, Washington

Jack Sullivan
University of Maryland
College Park, Maryland

Introduction

The Early Development of the Garden in Europe

This history of European gardens begins in fifteenth-century Italy—a time of rebirth, a period of awakening for an entire culture that for centuries had shut itself off from the outside world. It was the beginning of the Renaissance. From the fall of the Roman Empire, in a period known as the "Middle Ages," people's lives had been focused on survival and religion. Northern Europe had been in a state of chaos and uncertainty. Society at that time had searched for peace and found it in spiritual reflection and the promise of a better life in the hereafter. Gardens reflected this introspective culture. They were a sheltered place designed for prayer and reflection and void of worldly pleasures. The Renaissance influenced a great change in this culture. It was a major ideological movement that had a profound impact on society. In this new philosophy for living, people reconnected with the world around them. They no longer feared retribution for worldly achievements. They grew in confidence, believing that a loving God would reward them in the hereafter for making positive contributions to the present world. Individuals began to take pleasure in the natural world, a world they had denied for hundreds of years.

Renaissance philosophies and artistic ideals found expression in the creation of the Italian villa, where house, garden, and surrounding landscape were designed as one. These country estates became one of the highest forms of artistic expression and gained in significance as more and more wealthy and prominent families commissioned their building. The principles of Renaissance design informed subsequent styles in other countries and influenced many great works.

The Renaissance garden was, and still is, a paradigm for the art of spatial organization. It was guided by a logical body of rules that sought a balance between man and nature. Scale, symmetry, and proportion were manipulated in compositions to achieve that balance, which reflected the ordering of nature and thus, according to Renaissance theory, nature's beauty.

Toward the end of the sixteenth century, this simple beauty, derived through the skillful structuring of space, was redefined by human invention and contrived pleasures. The Renaissance garden of the sixteenth century became something quite magnificent. It was mostly the ambitions of an elite society of religious officers that affected the changes. Grottoes, fountains, and extraordinary sculpture contributed to an overwhelming display of exuberance surpassed only perhaps by some of the creations belonging to the seventeenth-century Italian baroque period.

Religion, and especially the Roman Catholic Church, gave rise to the distinctive baroque style of design that conveyed the power and dignity of the

Church by employing very dramatic design elements, rich in decoration and elaborate detail. By the second half of the seventeenth century, the baroque style had moved from Italy and emerged in France as the grand and elegant style of kings. Its magnificence glorified the power of the absolute monarchy and celebrated France's position as a leading nation in Western Europe. The seventeenth-century French style developed its own identity as a style of magnificent beauty and perfection. It flourished as a powerful political statement and represented the fine taste of a privileged culture whose wealth and ambition had no boundaries.

From the early decades of the eighteenth century, coinciding with the death of King Louis XIV, France's position as the dominant power in Western Europe began to dissolve. Under a new monarchy that no longer ruled the country uncontested, the privileged and celebratory style that had formerly characterized French gardens began to wane. As the monarchy's ability to run the nation grew increasingly questionable, and public doubt and opinion gave rise to revolution, this grand style became a static memento of royal absolutism. While grandiosity and lavish excesses had all but disappeared from new French gardens, the French style continued to show up in other European countries that had adopted its former principles to celebrate their own monarchies.

In England, French developments had inspired many great royal palaces and gardens since the beginning of the sixteenth century. France was positioning itself as the cultural leader in Europe, and its successes could not be ignored by its European contemporaries. But by the eighteenth century, the elements of England's adopted French style were gradually replaced by a new garden theory that was more attuned to nature. A change in English culture engendered a change in the idea of the garden. Art, literature, and politics all contributed to a new garden theory that celebrated nature's inherent beauty. The natural landscape became a model for garden design, thus gardens began to look as though they had naturally evolved.

In the early stages of development, these gardens were designed to represent the beauty of nature as it was depicted in classic scenes painted by popular seventeenth-century artists. Idyllic environments of the classical world were recreated in the English countryside. Monuments, temples, and statues were designed and added to these landscapes as part of the Arcadian scenery. In the garden, nature became a unique art form and a place for reflection.

In the next phase of this period, beginning in about the middle of the century, gardens became more natural looking. They were less symbolic or representational and tended to be more simply designed. The landscape was manipulated to create beautiful, scenic parks that provided aesthetic satisfaction to their owners as well as recreation and income. The parks were hunted and grazed, and the surrounding woodlands provided cover for game and yielded timber. It was this style that England became most known for, a style that would be copied not simply for its pleasing aesthetic effects but for the peace and tranquility that those effects provided to garden viewers. Later, gardens would become more untamed removing nearly every sign of human intervention.

During the nineteenth century, the Industrial Revolution dominated England. It yielded a new class of wealthy entrepreneurs that celebrated their successes by building lavish homes with exotic gardens. The English park-style landscape made popular during the eighteenth century was now replaced by a new eclectic fashion that took inspiration from past styles. In a frenzied attempt to represent the most appropriate aesthetic for this nouveau riche society, many period styles were copied and merged, resulting in a dizzying collection of ideas that often lacked coherence. There were, however, some positive aspects to Victorian gardens that made the period worthy of its place in garden history. A unique aspect of these gardens was the inclusion of many new and interesting varieties of exotic plants in design schemes. Plant collecting had grown popular with the wealthy society, and the garden became a place to display exotic collections from around the world. Exotic plants found their way into the garden either as part of the designed landscape or under the protection of glasshouses, also developed during this period to support the growing interest in cultivating foreign plants. The Victorian age, though never hailed as a period of great garden design, is recognized for the advancements made in horticulture breeding and technology. This progress played a major role in future garden development, especially in the horticulturally rich gardens that became popular at the beginning of the twentieth century.

Spurred on by the rising costs of land ownership and a new economy that was less dependent on agriculture, the wealthy classes began acquiring smaller properties in the early decades of the twentieth century. The country estate was still popular, in fact, even more so now, as advances in industry throughout England during the previous century made country life a very desirable, sought-after lifestyle. So much of the country's natural landscape had been raped to satisfy the needs of industrial progress that country estates and their gardens became havens to enjoy nature's inherent beauty. Gardens were bountiful and plants grew unhindered. The general shift of taste favored gardens that were designed to be lived in, places where one could escape to and find pleasure. This informal style of gardening continues, with similar rewards for those in our contemporary culture who seek the pleasures of the garden as an escape from the burdens and trials of today's world.

There is a rich history to be told in gardens, a history of the lives of those who owned them and of those who created and tended them. We can, with thoughtful interpretation, draw lessons from these accounts and gain valuable insight into the past cultures that inspired great historic gardens. Equipped with a better understanding of the ideas and methods that informed these gardens of the past, we begin to have a greater appreciation for them and are thus able to recognize their usefulness as a valuable resource and inspiration for contemporary garden design.

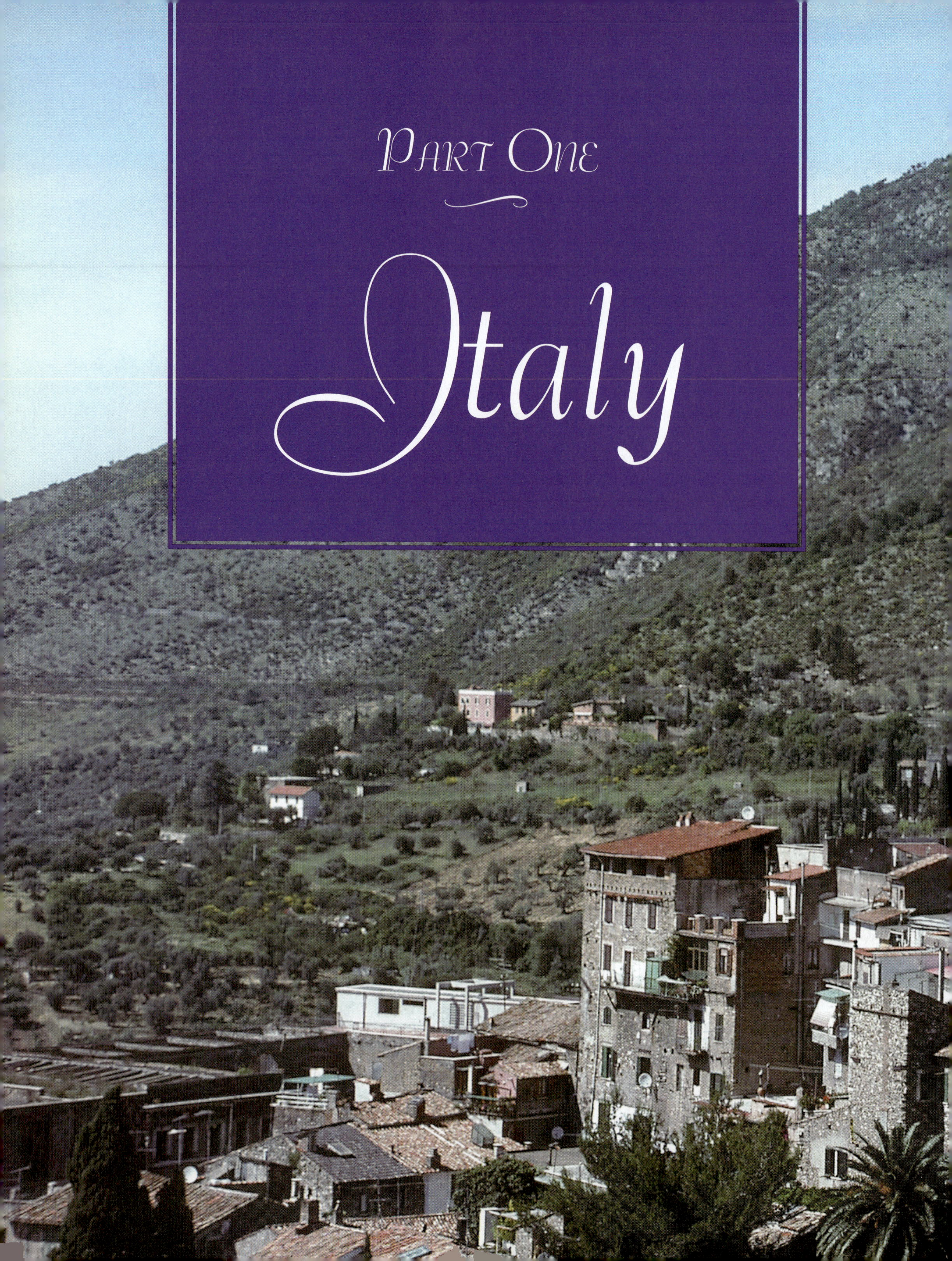

Part One

Italy

Environment

Italy's natural landscape is predominantly composed of mountains and rugged hills. To the north-northwest are the Alps, which extend to the Po valley from which the Apennine chain originates. The Apennines spread to the south from Genoa to Calabria and over to Sicily. In most areas along the Italian peninsula, the mountains extend down to the sea, which results in a very narrow beach or coastal plain.

Although the region receives adequate rainfall, averaging approximately 25 to 35 inches a year, rainfall distribution varies across the country. The Po valley experiences severe winters and warm summers, with heavy precipitation in the spring and fall. In contrast, the central and southern areas have a more Mediterranean-region climate, with approximately 80 percent of the rainfall occurring in the winter months. The landscape is accordingly quite dry in the summer, with rivers and streams greatly diminished.

Much of the history of Italy's designed landscapes, particularly the development of the Renaissance villas, occurred in Tuscany and in or near the city of Rome. In Tuscany, development occurred in and around the larger, prosperous cities, especially Florence. Newly designed villa complexes were home to the wealthy mercantile classes who established country residences high in the hills outside the cities to escape to during the hot summer months. The region around Florence was particularly desirable for the beauty of its natural landscape, shaped by the valley of the Arno River and elegantly enriched by the presence of orchards and vineyards covering the hillsides.

In Rome, fifteenth- and sixteenth-century villa construction was greatly influenced by a new era of Catholic reform. Church officers reinstated their authority in Renaissance culture by commissioning great building projects throughout the region. Building sites were selected across the Tiber river valley on a topography varying between steep slopes to the west and gentler rolling hills on the river's east bank. Development occurred mostly in two distinct areas. The first, southeast of Rome, was the region where most of the ancient Roman villas had been constructed. This area's history was both intriguing to the wealthy and inspirational to the artists and architects who were commissioned to build there. The second area of considerable development was northwest of Rome. This region had gained popularity during the Middle Ages as a favorite place for church dignitaries to establish their summer residences. It continued to be a desirable location for country estates during the Renaissance period.

By the middle of the sixteenth century, nearly all of the major cities north of Rome experienced some level of new development prompted by the spirit of pleasurable living associated with country estates.

Social and Cultural History

Business opportunities, politics, and religion were, until the age of the Renaissance, the primary incentives for travel to Italy. With the Renaissance came a new way of thinking, a new outlook on life influenced by a rediscovery of the natural world and of an individual's place in it. A new culture developed, finding expression in the arts, architecture, and literature. This great intellectual and artistic movement dominated Italy from the early fourteenth to the mid-sixteenth century, ultimately expanding into all of Europe. By the late seventeenth and early eighteenth century, people developed a new appreciation for its art, music, and architecture, as well as for its natural beauties.

Rome and the Empire

The Romans were the first true citizens of Italy, a civilization that brought it to the forefront as the first European country to become an organized state. Early Roman settlements ultimately united, organizing themselves into a town. Rome emerged as an independent republic in 509 B.C. and began expanding its holdings through a crusade of conquests, much of which occurred in the fourth century B.C. By A.D. 116, Rome controlled a vast empire that included most of Western Europe. Historians have attributed much of Rome's success to a sound understanding of political and social order, a government system that worked in accordance with an established body of law, and a superior army. In contrast, the Romans attributed all of their successes to the power of the gods; Rome was great because it had the favor and divine protection of the gods, primarily because its leaders acted in accordance with the gods' wishes.

With the ongoing threat of invasion and actual invasion, the Empire was forced to alter its governing strategies and become more defensive under Marcus Aurelius (A.D. 161–180). With this change of momentum came a decline in Rome's power. By A.D. 395, the Empire was split in an effort to reorganize its system and regain control. These efforts failed, and by A.D. 476, the Empire was lost.

Early Rome's Contribution to the Arts

Early Rome is hailed for its contributions to the arts. Two of its principal contributions were architecture and urban planning. Roman civilization was urban. It developed as a network of towns, carefully organized by engineers who were accomplished surveyors with an appreciation for design. Roman engineers also experimented with architectural form and various building techniques. Basic principles of proportion and unity derived from Greek precedents were the foundation of their designs. New materials, such as mortar and concrete, were developed, and innovative techniques for building, such as the barrel and groin vault, were invented. These advances created new opportunities for the design of buildings.

Rome's buildings were quite diverse. They included palaces, country villas, and urban residences. Massive structures were built for public use as well. The Circus Maximus, for example, created primarily to host chariot and horse races, was an enormous structure that accommodated over 150,000 spectators. The Colosseum, another spectacular construction, was designed as an amphitheater that held over 45,000 spectators. It was principally used for gladiator games and animal fights. Another major architectural accomplishment was the design of the public baths. These baths were part of a larger building complex that included many buildings of various public interest, such as libraries and museums.

The Middle Ages

The Middle Ages was a name assigned by Renaissance scholars to the period of time between the Fall of the Roman Empire (A.D. 476) and the Renaissance (fourteenth through sixteenth centuries). It was a period of interesting development in European culture, though most of the achievements of the middle centuries have been overshadowed by the advancement of Christian faith. The Roman papacy emerged as the spiritual head of the Western church. It was an age of religious dedication, of belief in the virtues of spiritual reflection, and in the beauty and peace bestowed on the faithful in the life hereafter. Absolute dedication to the future world relieved the fears of the present, fears derived from the chaotic world that ensued after the decline of the Empire.

Individuals in search of peace and a restored faith often escaped from society, turning their back on worldly pleasures and focusing inwardly on their own spiritual journey. Many sought out monastic communities, where prayer, discipline, and hard work captured one's mind and nourished one's soul. Early ideas for community planning can be observed in the organization of these communes. The physical organization of residents within a monastic community was accomplished through the design of a cloister. Monks were grouped into separate enclosures situated along a covered walkway designed as four equal paths that formed a complete square. The resulting interior space was an intimate courtyard, often planted as a garden. The gardens were utilitarian, planted mostly with medicinal herbs and, later, flowers. The flowers were not grown for simple pleasure but rather for altar decoration in their place of worship.

While the earliest gardens of this middle period seem to have been entirely practical with plantings of vegetables and herbs, evidence reveals that later gardens were somewhat more artistic. Garden scenes depicted in medieval manuscripts suggest that people gradually began to enjoy the garden simply for its beauty. Flowery meadows, bubbling fountains, and grass-carpeted garden seats replaced vegetable and herb plots. Yet these spaces, as pleasurable as they were, were still hidden behind walled enclosures. The garden reflected the medieval culture, inwardly focused and uninfluenced by worldly pleasures. Much like the medieval faith, it was a sanctuary of inner peace and hope.

The Renaissance in Italy

In the decades preceding the Renaissance, Europe's cultural reawakening, Italy was a politically fragmented country. Cities, feeling a common freedom from any outside threats, began creating their own municipal governing bodies. These independent governments were engaged in endless battles with each other. The country became even more fragmented as wealthy landowners sought to gain control of the districts where their goods were sold. In addition, powerful dynasties began to have a voice, applying family resources to gain their share of political control. In Rome, the popes maintained a modest level of power until 1316, when the Curia (a structure of administrative bureaus or departments created to assist the pope in his administration of the Church) was established at Avignon after the arrest of Pope Boniface VIII.

From the mid-fourteenth century, the wealthiest states—Milan, Venice, Naples, Rome, and Florence—came to dominate Italy. Florence, an especially prosperous business center, had become the wealthiest city in all of Europe by the mid-fourteenth century. The city earned its profits from the cloth industry, commerce, and banking. The profits from these businesses fueled the local banking industry, which grew to become an international system by the fifteenth century. The Medicis, a family of wealthy merchants and international bankers, seized control of the government in Florence in 1434. They ruled the city and the Tuscan region for over 300 years, beginning with Cosimo dei Medici, known as "Cosimo the Elder" (1389–1464). The city continued to prosper throughout the fifteenth century, absorbing a number of smaller cities but allowing them to maintain their individual identities.

It was during this period of the fifteenth century that a significant change occurred in the way society viewed the world. Individuals became less preoccupied with life in the hereafter and more appreciative of life in the present world. Old scholarly and artistic achievements were observed in a new context. People looked back at the accomplishments of the classical world, particularly of Greece and Rome, seeking to match or even to move beyond them. Scholars encouraged the revival of the classical past to support new learning associated with these ideas. This new philosophy was supported by a concept known as "humanism." Humanists encouraged the pursuit of a good life in this world, believing that "worldly" progress was perfectly compatible with, and even complementary to, Christian values. For Italians, reaching back to the past was restorative. It brought them closer to a time when they dominated the world, providing examples of excellence in politics, literature, and the arts. During the fifteenth and sixteenth centuries, the classical world was explored beyond its accomplishments in art and literature.

Classical architecture, town planning, and garden design were now being examined and drawn on as precedents for new building projects. Artists and architects continued to secure plenty of work throughout the period from an elite society that sought to express its dignity and status by commissioning great works of art or by constructing extravagant villa complexes.

This artistic movement spread to Rome in the early 1500s as a succession of ambitious popes and cardinals attempted to transform the city into something more than the seat of Catholicism. As artists turned their attention to Rome, they found a culture in the midst of change, one that began questioning the very ideals that fueled the Renaissance. The sacking of Rome in 1527 with the destruction of the churches is often noted as the event that seized the momentum of the Renaissance. While it did indeed shatter the faith of the humanistic culture, it was not the only cause for the change in artistic style that occurred "post Renaissance." Italian artists and writers were likely affected by the period's unrest. After all, people had begun questioning the ideas of humanist faith that had inspired these talented individuals for nearly two centuries. But artists and writers, it seems, were now seeking a new form of artistic expression, something fresh and energetic. What emerged was the pursuit of a new art form that celebrated individual accomplishments rather than those of an entire society or a culture. This new approach did not deny Renaissance precedents but rather looked to extend beyond the principles of those achievements to create something more elaborate and expressive. It was this new style in the arts that paved the way for the intense period of artistic activity called the "baroque," a derogatory term assigned by neoclassicists to the period of art beginning in the seventeenth century and lasting well into the first half of the eighteenth century. Beginning circa 1600, most baroque art was commissioned by the Catholic Church as a way to combat the advancement of Protestantism. Scenes were painted to capture and exploit the religious passion in people, thus encouraging or strengthening their religious association. Throughout the seventeenth century, baroque art and architecture essentially transformed the city of Rome. Through the contributions of the finest architects and artists, a poor, deteriorating town became one of the finest cities in all of Europe. The style soon became popular with the aristocracy, attracted to its overwhelming and ornate design techniques. The style was soon transported to great palaces all across Europe.

Chapter 1 Elements of Italian Garden Design

Figure 1–1 A vine-covered pergola at Villa Medici in Fiesole.

The Italian Renaissance villa garden was a significant part of the country experience that was popular among wealthy Italians during the fifteenth century. It evolved as the model for the developing Italian Renaissance style of design. The villa combined the relationship of house, garden, and countryside into a single composition. Terraces, **loggias** (a roofed porch, gallery, or arcade, typically attached to a building with an open colonnade on one or both sides), stone steps, and **pergolas** (a covered walkway, usually framed by a double row of stone or wooden posts with crossbeams above, which are typically covered with climbing plants, Figure 1–1) linked house to garden, while hillside prospects provided visual links to the natural environment that surrounded these country estates. The essence of the country residence was to experience the benefits of nature. Its location outside of the polluted cities, typically in the hills, provided fresh air and sunshine, both highly regarded as prerequisites for good health and comfortable living. Villa gardens were designed to encourage outdoor activity and to provide plenty of opportunity for recreation and exercise. The Villa Castello and the Villa Petraia, both located in the foothills of Monte Morello, were two of the earliest country estates built for the wealthy Medici family of Florence. Their hillside locations are typical of the sites chosen for country estates. The clean, dry air and gentle mountain breezes offered the perfect environment for a summer retreat.

Hillside sites required a designer's skill for creating a plan that would take advantage of the benefits of the location and its views and at the same time be sympathetic to the natural surroundings of which the property became a part. Careful attention to scale and proportion blended new development with the countryside. Garden terraces were carved into the hillsides in the same way that the ancients had manipulated these landscapes centuries before to accommodate the planting of vineyards and olive groves. The Villa Medici in Fiesole, one of the first Medici villas, is regarded as a fine example of Renaissance design. It is a spectacular achievement even by today's standards in its successful site selection, design, and development. At the Medici estate, loggias and pergolas link house to garden, and terraces offer views that connect garden and countryside. Spaces in the garden are carefully proportioned and beautifully blended into the natural hillside setting.

Social and political factors became increasingly important for villa development late in the century. The domestic quality that characterized the early villas was replaced by lavish displays of wealth and status. One of the finest examples is the Pitti Palace and its associated Boboli Gardens in Florence. They were used as a royal estate for the Medici family to celebrate its political power and prestige and to reassert its position as the leader of the most influential city and region in the country. The gardens, sized to accommodate a large number of people for ceremonial events and court festivities, today seem rather open and overwhelming in scale, but the spaces were well proportioned for the celebratory events that animated the garden during the Medici reign.

The design of the Renaissance villa was a product of several opportunities and constraints. Its specific site location was one of the first considerations. Beautiful vistas provided visual stimulation. In some areas, a view out over the adjacent city became a unique sort of status symbol for the wealthy families that politically controlled those precincts, thus it was important to carefully select its location. Climate also was a factor that influenced the design of a villa garden. Country homes typically were located on south-facing slopes, open to the sunshine and summer breezes. While this offered health benefits, it created uncomfortably warm conditions as temperatures climbed during the summer months. Certain features and materials incorporated into the design of the garden mitigated the heat's effects. The presence of water, primarily through fountains, provided a cooling effect, as did the shade from constructed elements such as loggias and **grottoes** (rustic, cavelike chambers decorated with rocks and shells and often containing fountains and other waterworks, Figure 1–2) built of stone. Trees planted singularly and in ***boscos*** (the Italian term for a wooded grove planted in the garden) also gave relief from the midday sun.

FIGURE 1–2 A stone grotto at Villa Gamberaia in Settignano.

Villa gardens functioned as the threshold between dwelling and natural countryside. They were poised between formality and informality, balancing the inclination for human control with the known benefits of yielding to nature and its inherent beauty. Gardens were designed in accordance with geometric principles that followed rules already established in architecture. Simple geometric forms, such as the square, the circle, and later the oval, were used to harmonize and reflect the forms established for the residence. These forms were developed as individual garden units or compartments, each designed to enrich the overall garden experience. These units, defined by walls, hedges, and pergolas, were organized symmetrically along a central axis that extended the length of the garden. Cross axes, often located at a change in garden level, further divided the composition and created access to other exciting features. Within each designed space were unique and exciting garden elements. Some were linked to a broader itinerary, while others existed as isolated, decorative components created simply for pleasure. It is this system of spatial organization that best characterizes the contributions of Renaissance design. Individual parts maintain their distinctiveness, and at the same time they are united with axes, vistas, symmetry, and the repetition of elements such as water, trees, and sculpture. Villa Gamberaia in Settignano displays nearly every quality of successful Italian garden design within the constraints of a relatively small site. Here the Renaissance principles of order and geometry are illustrated in a sequence of carefully proportioned spaces linked by axes and beautiful vistas. Villa Garzoni in Collodi is an example of a successful garden confined to a rather awkward site. The design of the garden takes full advantage of a rocky hillside, producing a succession of terraces with views opening up to a host of designed features and a beautiful country landscape. Elaborate staircases, water cascades, and colorful ***parterres*** (a level space usually adjacent to or near a building or house in which geometrical patterns are created from plants, best viewed from a prospect above the garden) add to the garden's unique and interesting shapes and patterns. Its baroque extravagance and ornamentation has an entertaining quality that draws one into the space and encourages one's progression through it.

Water played a significant role as a point of interest and also as an element to connect various spaces both visually and thematically. Hillside sites made it possible for water diverted from nearby rivers and streams to flow naturally through gardens feeding elaborate fountains and creating spectacular effects by employing the forces of gravity. At Villa d'Este in Tivoli, water is the main attraction. Spectacular fountains and *giocchi di acqua* (or water games, fountain jets designed to surprise unsuspecting visitors with their water spray) appear throughout the garden as part of a demonstration of the owner's superior taste and creative vision. Water also is a dominant feature in the garden at the Villa Lante in Bagnaia. There the entire garden scheme is organized around water, which takes on a variety of garden forms to create the garden's main axis. Fountains, cascades, and pools link the various garden compositions and function as each area's main element.

The "idea" of the villa garden, that is, its context or meaning, often was conveyed by the inclusion of sculpture. While many pieces were enjoyed simply for their beauty, most were appreciated for the symbolic message that they conveyed, which generally was part of a larger context that often directed the garden's overall theme. Statues of ancient gods and goddesses were placed in significant locations in the garden as individual elements or as part of elaborate fountains. Grottoes also housed ancient figures. These symbols of ancient mythology were known and understood by a Renaissance society that was fascinated with the accomplishments and virtues of the ancients. This symbolism is less known today, of course, but the sculptures are no less captivating. The Sacro Bosco at Bomarzo is perhaps one of the best examples of incorporating art into nature. Giant sculptural figures carved from native stone appear throughout the park. These massive sculptures likely alluded to a deeper meaning understood by a sixteenth-century culture, but today, the Sacro Bosco, known as Parco di Mostri, or monster park, is popular simply for its amazing collection of strange-looking figures, elephants, lions, bears, and dragons that magically greet unsuspecting visitors who venture into the dense forest.

Three basic categories of planting were consistent with most Renaissance gardens: the wooded groves, or *boscos,* orchards, and smaller flower and herb gardens. Plants were arranged in formal patterns in the garden, consistent with the overall site layout. Evergreen hedges of boxwood or laurel outlined garden spaces and defined paths (Figure 1–3). Rows of evenly spaced trees, such as evergreen oak and sycamore, reinforced axes and further defined garden divisions. Larger tree species such as these also were planted in the *boscos* along with several types of pines, such as umbrella and scotch pine. Fruit trees were quite common to the Renaissance garden. The pomegranate, apple, lemon, and bitter orange were most popular in early gardens. Nut trees also were very popular. Almonds, hazels, and walnuts were among the most commonly grown. Tender plants, such as citrus, were grown in the garden in ornamental pots that could be moved to the shelter of a *limonaia* (Lemon House) during the winter months. Climbing plants, such as honeysuckle, wisteria, grapes, and roses, covered walls and draped pergolas to focus or obscure views. The smallest garden spaces contained the greatest variety of plantings. Vegetables, herbs, and flowers were grown in individual compartments laid out in a geometric scheme. This area of the garden was called the

FIGURE 1–3 Evergreen hedges and topiary in a Tuscan garden.

FIGURE 1–4 A parterre *at Villa Lante in Bagnaia.*

orto, or the *giardino di compartimenti* (garden of compartments). A garden similar to a *compartimenti* in size and plant composition but more enclosed or private and located near the house was known as the *giardino segreto* (secret garden). This enclosed garden retained the spirit, in terms of its layout, not religious symbolism, of the medieval *hortus conclusus* (an enclosed medieval walled garden).

From the mid-sixteenth century, a new and rather elaborate planting design technique became popular. Geometric patterns were composed within compartmental squares and referred to as *parterres* (Figure 1–4). The patterns typically were created with low hedging plants, such as myrtle and privet, between which sand or gravel was spread on the ground plane. These intricately patterned beds were intended to be looked down upon from the house or from an elevated prospect such as an upper terrace. The French later adapted this technique, making it even more elaborate by copying patterns from fine embroidery and transferring the designs to gardens called ***parterres de broiderie*** (ornate patterns on the ground created with gravel, herbs, and evergreen plants such as boxwood or yew).

The Italian Renaissance design style dominated Europe throughout the sixteenth century. Its lessons informed many subsequent design approaches. Classic principles and formulas for organizing and manipulating space, combining forms, and working with scale and proportion were important for designers to know. These insights and approaches allowed them to create new and exciting landscapes that were informed and influenced by past successes but modified to suit their individual programs and tastes.

Chapter 2

Landscape Expression: The Renaissance Gardens of Tuscany

P*er semplice diletto* (simply for pleasure) were the words that fifteenth-century Italian architect and writer Leone Battista Alberti used to describe the villas and gardens of Renaissance Italy (Ackerman 1990). It was the peacefulness of life in the country and a love of nature that brought the Italians from the crowded cities to the rolling hills of Italy's beautiful countryside during the fifteenth and sixteenth centuries. This period coincides with the great revival of art and literature known as the "Renaissance."

The Renaissance brought about a new concept of the individual, a restored faith in human abilities, and an appreciation for each individual's role in the natural world. In addition to this human-centered spirit, the Renaissance prompted a renewed interest in the classical world. Renaissance humanists (those who focused on the capabilities and values of humans) looked to rediscover the values of antiquity, predominantly of the Greek and Roman civilizations, as the foundation for a new civilization where the intellectual could prosper and still be loyal to her or his faith. Classical texts prompted a reunion between humanity and nature. Writings such as those by Roman poet Virgil (70–19 B.C.) described the virtuous aspects of life in the country, where a tranquil style of living encouraged civilized behavior. Virgil spoke of the contemplative man who discovered the beauties of the natural world while restoring peace within himself. This perfect scene of the natural world, an earthly paradise, or Arcadia, as envisioned by Virgil, would be sought by a host of wealthy patrons during the Renaissance.

A supporter of the humanist movement was the wealthy banker and statesman Cosimo dei Medici, "the Elder" (1389–1464). Cosimo, seeking to encourage humanist theory and education, assisted humanist scholar Marcilio Ficino (1433–1499) in the founding of an academy called "Academia Platonica." Ficino's academy acquired a country estate called "Villa Careggi," near Florence, from Cosimo in 1462. Soon this center for learning became a gathering place for many great minds and influential people, including officers of the Catholic Church. The academy, inspired by Greek philosophy, developed as a place of contemplation and scholarship in the spirit of Plato (428–347 B.C.), the Greek philosopher and one of the most creative and influential thinkers in Western philosophy. Humanist scholars emulated Plato by making the garden an important component of institutional education. Plato was known to have enjoyed teaching in his garden, as he considered it one of the highest forms of artistic expression.

From the end of the fourteenth century, the art of garden design found its expression in the creation of the villa, or country house. It is important to note here that the term *villa* refers to the entire estate, both the house and surrounding landscape. The idea of a country residence dates back to the Romans, who built villas in the country to escape life in the city. Guided by ancient building practices, Renaissance architects developed their own theories for architectural design and garden planning. The villa and garden were conceived of as one artful composition. Indoor and outdoor spaces were designed together, with a visual dialogue between them that encouraged the unity of spaces. In his influential treatise on architecture, *De re aedificaturia,* published in 1485, Alberti (1404–1472) (Enge, Enge and Schröer 1992) made clear the rules of classical architecture, interpreting the essence of the classical language for a new generation of designers. He concluded that beauty and, accordingly, nature's beauty result from a perfect harmony of all parts within a composition. On the basis of Alberti's theories, designed nature (the garden) was to reflect the natural proportions by which true nature exists. His ideas, specific to villa design, focused first on the site, as appropriate siting was paramount to success. Alberti recommended that villas be located on hilly sites with peaceful views of surrounding countryside and exposed to the sun and wind (both of which contributed to the healthfulness of country life).

The garden was designed together with the house and thus followed similar rules of proportion and geometry. Garden terraces, loggias, and stairways were to provide links between the house and garden. Terraces further joined the garden to the neighboring countryside, with views of the distant landscape. The ideas of Alberti and other artists, writers, and philosophers were shaping a new culture that revived from extinction some of the great successes in the arts of antiquity.

The construction of villas in Renaissance Italy beginning in the fifteenth century coincides with the growing prosperity of the successful merchant families of Florence. These wealthy business families often owned more than one home. They typically had a house in the city near their business and one in the country. Their country estates were a pleasant retreat from city life as well as an investment. Many estates began as working farms, generating a fair amount of income, but by the end of the fourteenth century, farming became less desirable. Successful entrepreneurs with considerable wealth sought to transform their working country properties into pleasure estates designed strictly for leisure, recreation, and entertainment. They commissioned some of the most highly acclaimed Renaissance artists and architects, including architects Michelozzo, Sangallo, and artists Pontormo and Bronzino.

Florence gained recognition as a center of culture, attracting many young artists and intellects. The city continued to grow under the auspices of the wealthy mercantile and financial aristocracy that essentially controlled the city's governing bodies. Despite frequent revolts by the democratic faction against the ruling class, the government continued to be strongly influenced by the wealth and power of the business class.

The control of the government was ultimately seized by one successful merchant family, the Medici family of Florence. In 1434, Cosimo dei Medici gained control of the Florentine government, which the family would rule for over three centuries. Upon acquiring this new position, Cosimo embarked on a series of building projects, both in and near Florence. Fond of the idea of a country residence, he set out to build several villas for himself and members of his family. Some of the earliest villas built or acquired by the Medici family included Villa Medici in Fiesole, Villa Castello in Florence, and Villa Petraia, also in Florence.

Designers, guided by the theories expressed by Alberti, began to lay out these villas based on developing Renaissance principles. According to Alberti, a relationship should exist between house, garden, and countryside. He advised: "I do not think it necessary for the Gentleman's house to stand in the most fruitful part of his whole estate, but rather in the most honorable, where he can, uncontrolled, enjoy all the pleasures and conveniences of air, sun, and fine prospects" (Bajard and Bencini 1993, 12). In terms of these ideals, Villa Medici was most successful. It was selected for its relationship to the countryside and designed to take full advantage of the site. The Medici family continued building and restoring villas into the sixteenth century. Nearly all of the Medici villas in and around Florence were built or redesigned for the enjoyment of the family, with the exception of the Boboli Gardens and the associated Pitti Palace, which became a symbol of the Medici dynasty as a whole. An inventory of the Medici homes was documented in a series of paintings by Giusto Utens in 1599. Utens, a Flemish painter, was commissioned by Grand Duke Ferdinando I dei Medici, Cosimo I's son, to document all of the villas belonging to the Medici family in a series of lunette paintings. These paintings, now on public display at the Museo Storico Topografico in Florence, reflect the original intent of the design for these properties, which is a valuable resource to garden historians.

Villa Medici Fiesole

1. Green Garden
2. Gateway
3. Small Terrace
4. Service Building
5. The Casino
6. Pergola
7. Arrival Terrace
8. Retaining Wall
9. Limonaia
10. Garden House
11. Coach House/Service Building
12. Lowest Terrace
13. Olive Orchard

Villa Medici
FIESOLE, ITALY

Villa Medici Time Line

1458–1461	The villa (house and gardens) is built by Italian Renaissance architect and sculptor Michelozzo di Bartolommeo for Giovanni dei Medici, son of Cosimo dei Medici, "the Elder."
1671	The Del Sera family purchases the property from Cosimo III.
1772	The villa is purchased by Lady Margaret Orford, sister-in-law of eighteenth-century English novelist Horace Walpole.
1915	Lady Sybil Cutting, then owner of the estate, commissions architects Geoffrey Scott and Cecil Pinsent to revise the west terrace and redesign the lower garden terrace.
1959	The villa is purchased by Aldo Mazzini.
2000	The villa, under new ownership, undergoes restoration work. It remains open for public visits by appointment.

Villa Medici, set in the hills five miles above the city of Florence, is unique in that it was developed as one of the first villas in Tuscany, where economic value was not the guiding principle for its creation. Before Villa Medici, country properties had historically existed as self-sustaining agricultural estates, where urban dwellers could go to escape the stresses and pace of the city and be cleansed or restored by directly interacting with nature. At Villa Medici, life in the country was completely separated from an agricultural context. The land, which is severely sloped, was not suitable for farming, suggesting that the location was valuable for other attributes, namely, its hillside prospect over the city of Florence and the neighboring countryside (Figure 2–1). With its sweeping view of Italy's finest countryside, Villa Medici anticipated a developing awareness of the aesthetic and restorative benefits of the natural landscape that would be expressed in painting, literature, and gardens for centuries to come.

The villa was built by Italian Renaissance architect and sculptor Michelozzo (1396–1472) between 1458 and 1461 for Giovanni dei Medici, Cosmo dei Medici's second son. Records indicate that Giovanni

FIGURE 2–1 View out to the neighboring countryside and the city of Florence.

FIGURE 2–2 Terra-cotta pots filled with citrus line the approach to the house at the arrival terrace.

closely supervised the construction of the villa and the garden. Giovanni died soon after its completion, leaving the villa to his nephew Lorenzo, Il Magnifico (1449–1492), who used the estate as a retreat for humanist philosophers.

The house and garden terraces of the villa are skillfully designed into a steep hillside with a commanding view of the capital city, perhaps the villa's most compelling feature. The complex is entered by an obscure gate tucked neatly into the landscape alongside a very narrow road that climbs the hillside to the city center of Fiesole. A narrow, shady drive beyond the gate opens to a large, rectilinear arrival terrace (Figure 2–2). This main terrace is a neatly designed space bound by a wall on the south side over which a hazy panorama of the Arno valley is visible. Terra-cotta pots filled with citrus flank the drive, reinforcing its linearity while adding interest to the approach through repetitive placement. Similarly on the south wall, a long row of potted red pelargonium creates an attractive line of color that directs one's view toward the neighboring countryside.

The casino, or villa house, is rather simple and straightforward, typical of Tuscan-style architecture. Its clear, geometric form and smooth, cream-colored exterior finish is brightened by its southern exposure, making it stand out against the surrounding vegetation. The placement of the house is a fine example of the qualities that Alberti envisioned in his treatise, which was completed at approximately the same time Villa Medici was being designed. Alberti's recommendations suggested that a villa should be located on a height to make its forms more grand, and with a garden and light all around it (Ackerman 1990, 76). The Medici villa opens to the garden, with commanding views of the neighboring countryside which, in essence, become part of the garden space. The importance of the villa's available views is reinforced in the design of various garden spaces where one is able to survey the distant landscape from multiple terraces offering a variety of perspectives. These carefully proportioned spaces, all relatively equal in volume, are carved into the hillside and supported by massive retaining walls that step down the sloped terrain. Through careful consideration of the existing site condition and a successful balancing of spaces, an extraordinarily perfect fit between constructed elements and existing landform is accomplished (Figure 2–3).

FIGURE 2–3 Spaces are carefully proportioned and beautifully blended into the hillside.

FIGURE 2–4 A vine-covered pergola below the arrival terrace reduces the scale of a massive retaining wall.

Just below the arrival terrace, on the next garden level, is a classic Italian pergola. The pergola is draped with roses and grapes that mingle beautifully in the overhead structure. Its placement along the high retaining wall successfully reduces the scale of the wall, which otherwise would appear massive and overpowering, and further provides a shady place from which to enjoy the distant views over the valley to Florence (Figure 2–4). Below the arbor, on the lowest garden level, is a simply designed garden dating from the early 1900s. It is composed of evergreens arranged in symmetrical patterns. This garden replaced an earlier and badly overgrown composition comprised of numerous exotic plants collected during the nineteenth century. The simplicity of the concept for this space is an effective foreground to the extended vista. A busier scheme would detract from the spectacular views provided from the overhead prospects, views that are, without question, some of the finest in any of the Renaissance gardens developed during the period.

The Del Sera family purchased the property from Cosimo III (1642–1723) in 1671. In 1772, the villa was purchased by Lady Margaret Orford, sister-in-law of eighteenth-century English novelist Horace Walpole. She made some changes to the villa and constructed the entrance drive, the Belvedere, and the Lemon House. In 1915, Lady Sybil Cutting, then owner of the estate, commissioned architects Geoffrey Scott and Cecil Pinsent to revise the west terrace and redesign the lower garden terrace, which had become quite overgrown with "exotic" plants. The villa today is still privately owned, but the gardens, fortunately, are open to visitors by special request.

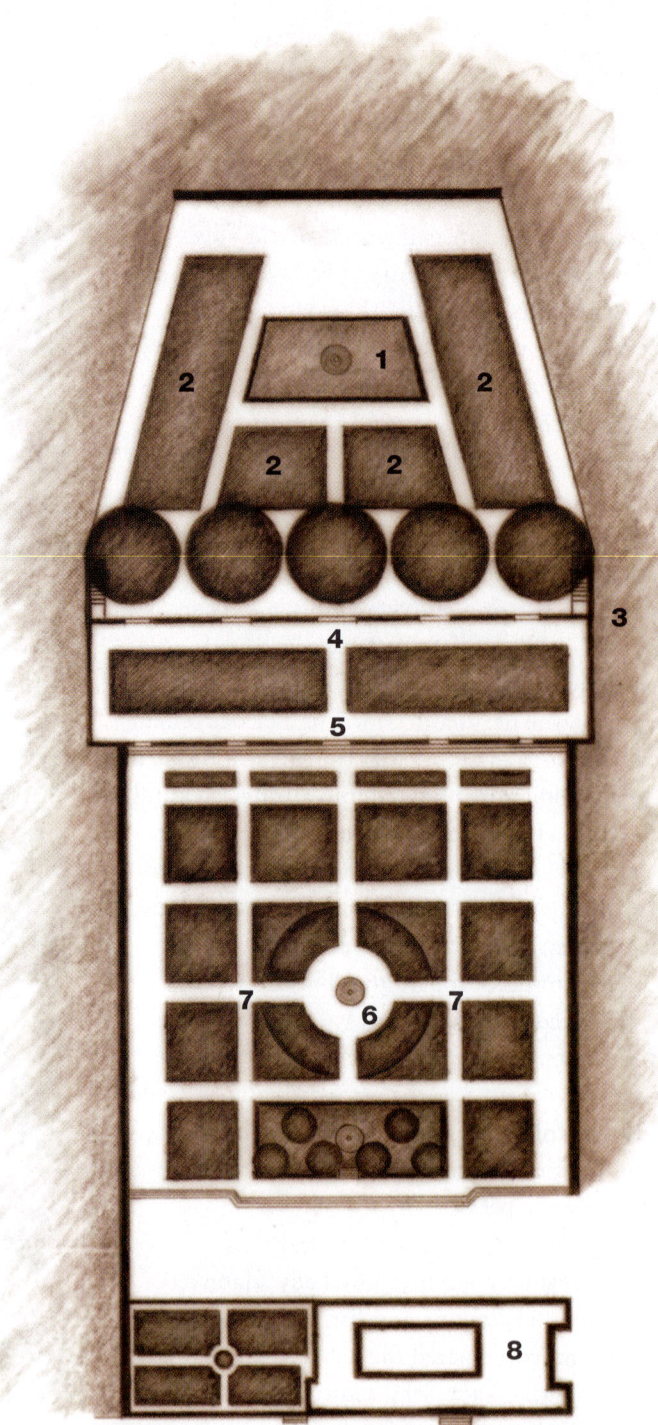

Villa Castello

1. Appennine Sculpture
2. Boscos
3. North Wall
4. Grotta degli Animali
5. Second Terrace
6. White Octaganal Fountain
7. Main Terrace
8. Casino

Villa Castello
FLORENCE, ITALY

Upon being elected Duke of Florence in January 1537, Cosimo I dei Medici (1519–1574) began transforming his country home at Castello into a grand estate commensurate to his new status as head of the government. The Castello property, inherited from his father, Giovanni dei Medici, is located northwest of Florence on the slopes of Monte Morello.

Cosimo was raised at Castello, which partly accounts for this being one of his favorite retreats. He returned often, especially in times of ill health, as the fresh country air was said to quicken one's recovery from illness. Excursions to the countryside during the fifteenth and sixteenth centuries were said to be restorative to both mind and body. It was believed that those who spent time in the country often were healthier than those living in the cities, as urban environments were quite dirty and unsanitary compared to today's standards. It was a privilege to escape the city environment, and that privilege belonged to the wealthy and politically connected, such as the Medici family. For the Medicis, the foothills of Monte Morello offered ideal conditions for a health-giving environment. The cool, dry atmosphere and pleasant breezes were ideally suited for those seeking a healthful resort (Wright 1996).

One of the main achievements in the development of Villa Castello was the acquisition of water, which is celebrated throughout the garden. It was acquired through the construction of an aqueduct in 1537 that transported it from the nearby spring to Castellina. The following year, Nicolò Pericoli (1500–1558), called "Il Tribolo", Cosimo I's favorite artist, was hired to design lavish gardens to surround the house. These gardens were to accommodate the many anticipated events that the family would host as part of its political status. In Tribolo's design, the site is organized around a central axis with three equally proportioned terraces carved into the hillside. The main garden, referred to as the "Great Garden," slopes upward toward the north end of the site. It is composed of a grid of square *parterres* created with boxwood hedges. These patterns are arranged symmetrically on either side of a central walk. Potted flowers and citrus punctuate the space, adding color to this otherwise green garden. The central path follows a subtle incline to the north, which offered the Duke of Florence a bit of modest exercise (Figure 2–5).

A fountain featuring sculptural figures of Hercules and Antaeus stood on the lower level of the main garden. These mythological symbols of strength and courage were an allegory of the Medici governance. In the battle between these two mythological powers, Hercules conquers the invincible

FIGURE 2–5 The main garden is composed of a series of organized parterres, *accented by large terra-cotta pots filled with flowers and citrus plants.*

Villa Castello Time Line

1537	Cosimo dei Medici begins transforming the grounds of his country home at Castello into a grand estate commensurate to his new status as head of the government of Florence.
	An aqueduct is constructed to bring water from the nearby mountains to the garden.
1538	Nicolò Pericoli, an accomplished sculptor and engineer, is hired to design gardens to complement the house.
1759	The estate, under new ownership, is significantly altered.
1780	The gardens are renovated in a neoclassical style.
1830	The wooded grove on the uppermost level of the garden is redesigned.
1999	Ammanati's Hercules and Antaeus sculptural group fountain is restored.

FIGURE 2–6 The Castello lunette by Giusto Utens (16th c.) is a fairly accurate record of the original garden scheme. Courtesy of Fototeca dei Musei Comunali di Firenze.

Antaeus. Cosimo likened himself to Hercules, who was portrayed as bold and virtuous, the qualities that Cosimo himself possessed, which helped him win over the anti-Medici party and lead Florence toward its position as the capital of Western culture.

Directly above this area of the main garden stood an evergreen bower, or a *labertino.* Once central to the main garden, it served as a focal point as well as a shady retreat. Within its evergreen boundary, one could escape the midday sun and be serenaded by songbirds perched in the trees. Centered in this space was the Fiorenza fountain, with a bronze sculpture of Venus, the goddess of love. This fountain, with its location among the evergreens, seemed to symbolize the rich, fertile land of Florence with its abundant supply of water from local rivers and springs. The fountain was later transferred to Villa Petraia, and the Hercules and Antaeus fountain was moved to the upper level of the main garden to take its place.

A lunette of Castello villa by Utens, the Flemish artist commissioned to record the family's collection of homes in a series of paintings, is our main resource for documenting the vegetation in the garden at the end of the century (Figure 2–6). At that time, fruit trees, both the regular and dwarf varieties, were planted as small orchards. Potted lemon and orange trees were arranged in the garden to provide both a pleasant fragrance and refreshing fruit for the table. These plants were sheltered from the weather by an **Orangery** (a garden building constructed with large windows on its south side to gather sunlight, used for overwintering citrus and other tender plants) to the west and a Lemon House to the east, each of which housed the potted citrus plants during the winter season.

A wall to the north marks the end of the main garden level. Along the wall are three large niches, the Grotta Degli Animali (Italian for "grottoes of the animals"). Each grotto contains sculptural groups of bronze and marble animals huddled together in stone basins. These animals seem to represent wild, untamed nature, here peacefully gathered and benefiting from a land rich in natural resources, such as water, which streams down the grotto wall behind them (Figure 2–7).

Beyond the wall, several steps lead up to a wooded grove. This is the uppermost level of the garden. A path winds through the trees to a basin that holds a large bronze sculpture called Appennino, created by Bartolomeo Ammanati in 1565. The sculpture is set upon a natural-looking rock. It depicts an old man in a shivering pose, the personification of the Appenine Mountains, the source of the great Florentine rivers' waters (Figure 2–8).

The Castello garden today appears as it did after the changes that occurred when the Medici dynasty died out. The *labertino* is gone, the fishponds in front of the palaces have long since been filled in, and the Venus fountain no longer stands in the garden, having been relocated to the Petraia villa. It is still a peaceful place, but it no longer has the spirit of greatness that one can imagine it had during the Grand Duke Cosimo's reign. Many of the changes followed a new fashion of design that developed during the late 1700s. In 1759, the estate, under new ownership, was significantly altered. Many of the trees were removed to open views, which essentially changed the entire composition of the garden. In the original design, the garden was laid out as a series of "outdoor rooms" extending from the house. Spaces were carefully proportioned and organized as a sequence of contrasting experiences faithful to Renaissance principles. Many of these interesting contrasts were lost as the garden became more open (Lazzaro 1990). Villa Castello provides important insight for today's designer as an example of a garden substantially changed by later revisions.

FIGURE 2–7 One of three large niches, called the Grotta degli Animali, or "animal grotto," carved into the north wall of the garden.

FIGURE 2–8 The bronze sculpture by Ammanati depicting an old man in a shivering pose is said to be a personification of the Apennine Mountains—the source of the great Florentine rivers.

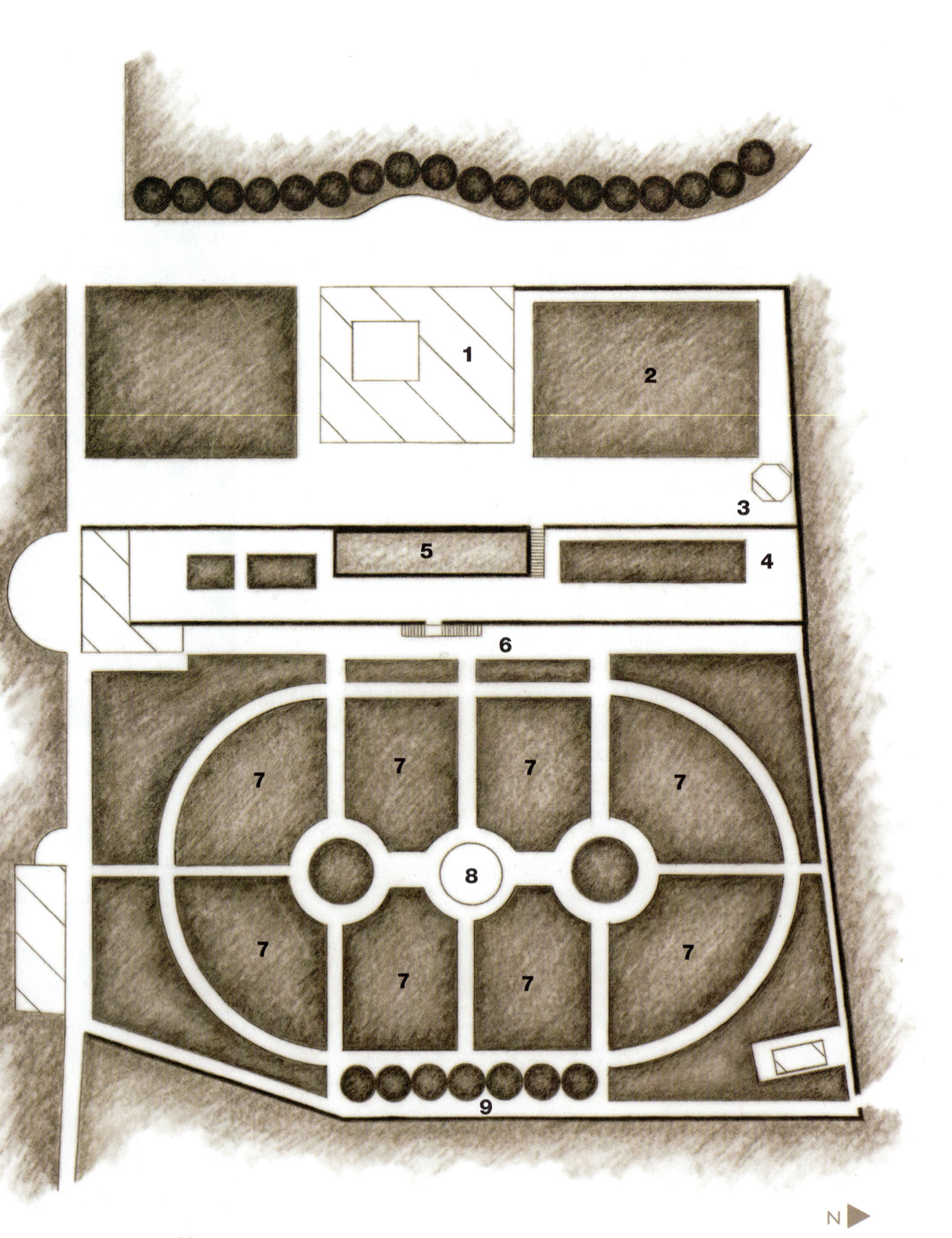

Villa Petraia

1. Casino
2. Garden of the Figurine
3. Belevedere
4. Garden of Palms
5. Fishpond
6. Double Staircase to Fishpond
7. Parterres
8. Mound with 3-Tiered Fountain
9. Plane Trees (platanus orientalis)

Villa Petraia

Florence, Italy

Villa Petraia, one of the early Medici pleasure villas, was added to the collection of Medici landholdings in 1532. It is located northwest of Florence, near Villa Castello. Villa Petraia, confiscated from the Strozzi family, did not become home to the Medici family until 1568, when Ferdinando dei Medici (1549–1609) decided to make it his primary residence. Ferdinando succeeded his brother Francesco as Grand Duke of Tuscany after his brother's death in 1587. Duke Ferdinando was a well-respected leader, with a very practical political mind, although this pragmatism did not extend to his personal life. He displayed a fond taste for the extravagance commonly associated with political status. Upon becoming duke, he set out to transform the Petraia property, then a working farm, into a magnificent villa with splendid gardens for his pleasure.

Work on the old fortified villa had begun rather slowly in the mid-1500s, as Cosimo's attention was focused on building the great city palace and associated gardens of Boboli. Nicolò Pericoli (known as "Il Tribolo") began the original garden layout at Petraia. Upon his death in 1550, the work passed into the hands of his son-in-law, David Fortini. It was not until 1588 that the property was transformed into the elegant Medici residence depicted in Utens's lunette of 1599.

Evidence from numerous documents disputes the popular belief that Bernardo Buontalenti was responsible for the elegant gardens produced toward the end of the century. It is believed that Buontalenti was involved in the restoration of the palace, but the creation of the gardens is largely attributed to Raffaello Pagni, a gardener from Florence. The palace, a thirteenth-century castle, was redesigned, but some of its medieval-style architecture, particularly the central watchtower, was left undisturbed. It is possible that the tower's fortress-like appearance was favored because it portrayed the villa as a protected retreat in the event that someone considered an attack in opposition to Medici rule.

The gardens depicted in the Giusto Utens painting from 1599 were organized around three terraces, the lower garden, originally referred to as *dei parterres,* dedicated to pleasure and growing fruit, the intermediate level, occupied by the large fishponds and small gardens on either side, and the upper level, planted with small trees, most likely citrus. There was a modest display of fountains and sculpture, since much of the water from the spring behind Petraia was diverted to Castello. The garden was laid out symmetrically with similar gardens on either side of the palace and central pathway. On the

Villa Petraia Time Line

1532	Villa Petraia is added to the collection of Medici landholdings.
Mid 1500's	The original garden layout is designed by Nicolò Pericolo.
1550	Nicolò Pericoli (known as "Il Tribolo") dies. The design and construction of the original gardens are passed to Tribolo's son-in-law, David Fortini.
1568	Ferdinando dei Medici takes up residence at Villa Petraia.
1588	Villa Petraia is transformed from an agricultural business center to an elegant residence. Gardens are completed by Raffaello Pagni, a gardener from Florence.
1805	The gardens are renovated, losing much of their earlier composition. Plane trees are planted along the lower boundary, a circle mound with a fountain is constructed on the central path, and the circle pergolas on the lower terrace are removed.
Late 19th c.	A stone belvedere is added to the upper terrace.
1919	The garden, owned by the Italian state, opens to the public.
Late 20th c.	The garden is replanted to recreate the historic period style. New fruit trees are planted in the orchards, and a collection of old bulb varieties is planted in the gardens.

Figure 2–9 The large fish pool on the second terrace.

lowest terrace was a secondary path crossing the main axis, joining two large squares, each a separate garden space. Each garden was defined by two concentric circular-shaped pergolas, one marking the garden's boundary and the other its center. Within the gardens, trees were planted in a ringlike pattern that followed the shape of the pergola ring. The second terrace, then and today, holds one of the most elegant features of the garden, a large fish pool created to supply fish to the villa. It is bridged by an elegant staircase (Figure 2–9) decorated with an intricate wrought iron railing. It ascends the hill to the casino level. On either side of the fishponds, flowers and herbs were grown in neatly organized plots punctuated with potted citrus plants. On the highest terrace, on either side of the palace, are small trees, probably citrus or some other type of small fruit. The prospect of the garden was the finest from this level, providing the best view of the neighboring countryside. The qualities of Renaissance design, perfect order, symmetry, and spatial harmony, were keenly displayed from this higher vantage point.

Many of these qualities of Renaissance design are essentially lost today as a result of nineteenth-century renovations. Plane trees line the garden's lower boundary, obscuring what were once beautiful views (Figure 2–10). A circle mound with a multitiered fountain interrupts the garden's main axis (Figure 2–11), and the lowest garden terrace is mostly open and uninteresting now without the circle pergolas that once dominated this area and balanced the composition.

What, then, is the value of this garden as an example of Renaissance style? Absent are nearly all of the original features that made it worthy of its place as a fine Renaissance garden, and that is precisely the justification for its study here. Villa Petraia today clearly exhibits the results of a lack of understanding of the original idea of the garden. The intent of the original design

Figure 2–10 A row of plane trees at the garden's lower boundary obscures garden views of the neighboring countryside.

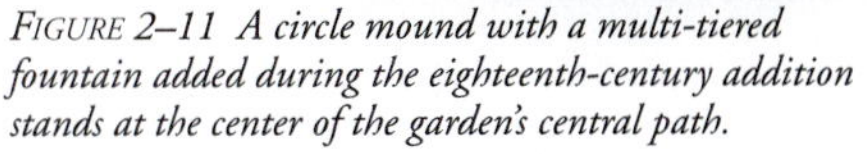

FIGURE 2–11 A circle mound with a multi-tiered fountain added during the eighteenth-century addition stands at the center of the garden's central path.

was ignored in later revisions. A comparison of the garden depicted in Utens's lunette (Figure 2–12) and the garden in its current state makes clear the importance of certain elements in the garden that helped characterize it as a Renaissance construction—balance, symmetry, compartmentalization of space, repetition, and a combination of intimacy and openness, most of which was eliminated from the garden in later "improvements."

FIGURE 2–12 The garden's original layout with a pair of circle pergolas situated on either side of the central axis depicted in Giusto Utens's lunette (16th c). Courtesy of Fototeca dei Musei Comunali di Firenze.

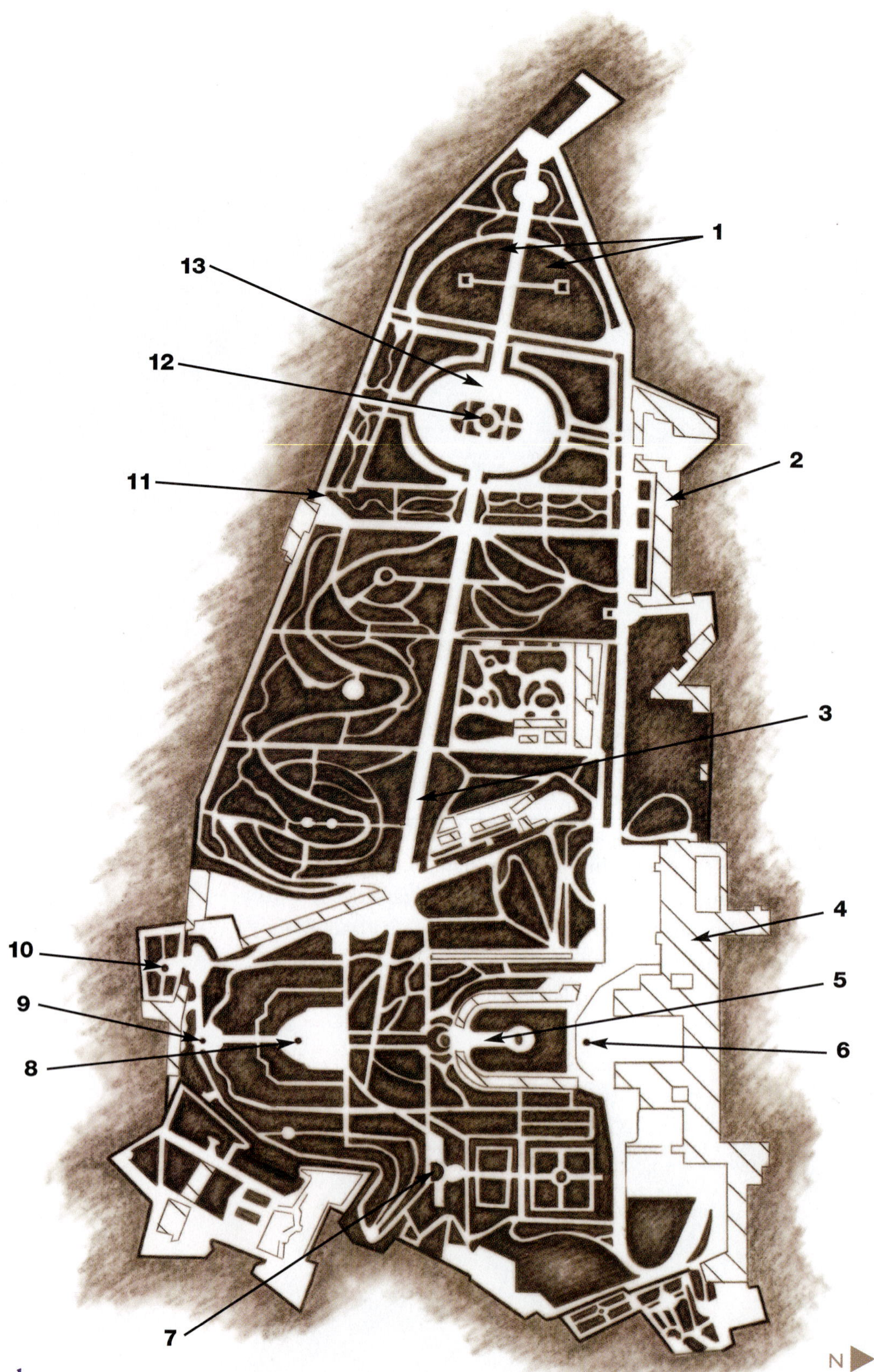

Boboli Gardens

1. Meadow of Columns
2. Lemon House
3. Cypress Lane
4. Pitti Palace
5. Amphitheater
6. Artichoke Fountain
7. Kaffeehaus
8. Forcone Basin
9. Statue of Abundance
10. Knight's Garden
11. Fountain of the Mostaccini (Moustache)
12. Isolotto
13. Island Pond

Boboli Gardens

FLORENCE, ITALY

The Boboli Garden complex extends south of Palazzo Pitti on the slope of the Boboli hill. These grounds on the left bank of the Arno River had been known from the late Middle Ages as "Boboli," a toponym indicating the woods that once thrived on the southern hill. Luca Pitti (1395–1472), son of an immensely rich merchant family, is remembered for building a palatial city residence on the site.

Luca commissioned Filippo Brunelleschi (1377–1446) to design a quattrocento palazzo, which would become, under the Medici family, one of the most splendid courts in Europe. A square, modular palace of the size that he had in mind would occupy an entire Florentine block. Taking advantage of the favor of Cosimo I dei Medici, Luca ordered numerous houses to be demolished along the Via Romana. To introduce a sense of distance from the Via Romana, Brunelleschi designed a palazzo fronted by a piazza. Elenora of Toledo (1522–1562), wife of Duke Cosimo I dei Medici, bought Palazzo Pitti in 1550. The purchase was motivated by the Duchy Court's need for a royal estate that was commensurate to its status and similar to other residences already established across Europe. Cosimo I reconceived the Boboli Gardens and Palazzo Pitti as a public setting. The grassy amphitheater south of the palace served as a vast celebratory stage set and also as a venue for court festivities (Pizzoni 1999).

The resulting landscape and architecture transcend the inspiration of Cosimo I. Rather than becoming a tribute to the vision of one individual, the common purpose of the *villa suburbane* (suburban villa), Boboli Gardens and Palazzo Pitti became a symbol of the Medici dynasty, flaunting Medician grandeur and prestige. The garden and palace, an extraordinary accomplishment, unparalled in scale, set the Medici family apart from the rest of society and clearly represented the family's political power and authority to an unprecedented degree.

Cosimo I envisioned a palace made more regal in the natural setting of the hill and valley. To transform the Pitti orchard (Orto de Pitti) into a formal garden, he commissioned Nicolò Pericoli (1500–1550), called "Il Tribolo," his favorite artist and the creator of many of the Medici gardens. Tribolo ingeniously made use of the predetermined site. For the main structure of the original garden he developed a U-shaped area, the amphitheater, which was an adaptation of a natural hollow (Figure 2–13). This hollow was part of an old stone quarry that had provided the stone used in the building of the palace. Tribolo also directed particular effort to create a system of drainage channels to prevent soil erosion on Boboli's steep terrain.

Water, the principal ornament and unifying element in the garden, originated from a spring in the Arcetri hills near Boboli. It was transported by aqueduct to the site and then across the Arno River to the Palazzo Vecchio

Boboli Gardens Time Line

1453	Luca Pitti commissions Filippo Brunelleschi to design a quattrocento palace (the Pitti Palace).
1550	Elenora of Toledo, wife of Duke Cosimo I dei Medici, buys Palazzo Pitti and the associated property. Cosimo I commissions Nicolò Pericoli, called "Il Tribolo," to design the gardens.
1560	The Cosimo I dei Medici family takes up residence at Boboli.
1565	A second aqueduct is constructed to carry water from the Ginerva spring to Boboli and also to the city, making it available to the population through fountains.
1612	Cosimo II commissions the Giardino del Cavaliere near the city wall, forming the southern boundary of the garden.
1620–1640	Alfonso Parigi is commissioned by Cosimo II to extend the garden westward, beyond its original boundary (the city wall of 1544). Parigi creates the Viottolone (Cypress Lane), and at its end, the Garden of the Isolotto.
1737	The last of the Medici males, Gian Gastone, dies. The Archduchy of Tuscany passes to the Habsburg-Lorraine.
Late 1700s	Leopoldo II replants and restores the gardens.
1799–1814	The Grand Duchess, Elisa Baciocchi, attempts to turn Boboli into an English-style garden, merging a very natural style with the garden's rigorously formal setting.
1834	Three large labyrinths dating from the 1700s are destroyed under Grand Duke Leopoldo II in 1833 to allow for a new serpentine carriage path.
1919	Victor Emmanuel III presents the palazzo and the associated gardens to the Italian state. The government renovates the gardens and opens them to the public.
20th c.	Theatrical performances are held in the gardens beginning in the 1930s.

FIGURE 2–13 The original Boboli layout, before its western expansion, shown in a sixteenth-century lunette painted by Giusto Utens. Courtesy of Fototeca dei Musei Comunali di Firenze.

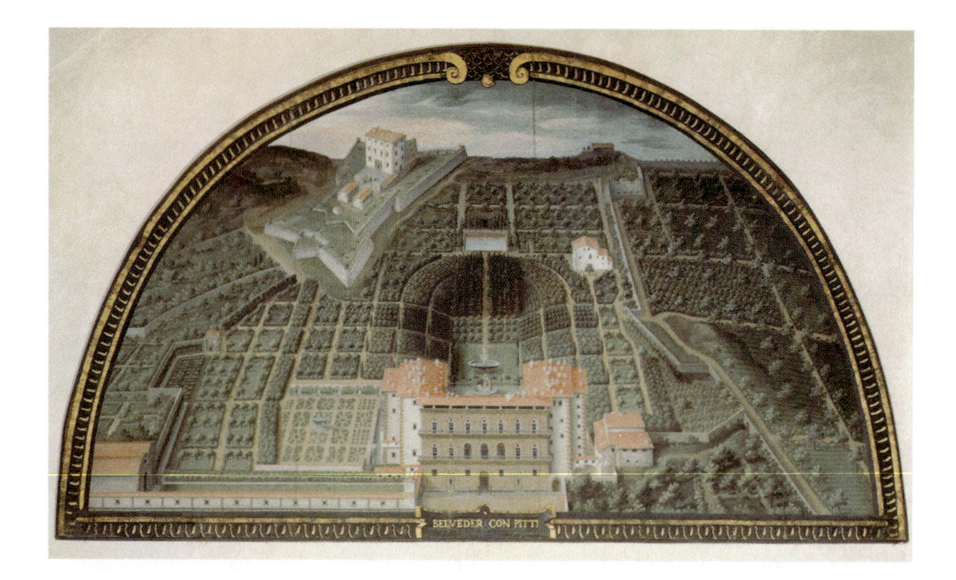

(old palace). A second aqueduct was constructed in 1565 to carry water from the Ginerva spring. Besides supplying the garden, both aqueducts brought fresh water to the city, making it available to the population through fountains, another example of Medici benevolence. Within the garden, Tribolo collected the water in a great basin beyond the amphitheater. He then redirected it through a system of underground pipes to other basins and spectacular fountains created throughout the garden.

Tribolo's design laid out the garden along a central axis, symmetrically aligned with the Palazzo and the hill (Figure 2–14). The line starts at the Fontana del Carciofo (artichoke fountain), near the Palazzo, and rises up the hill, from the northwest to the southeast. Below the fountain is the amphitheater, designed to host ceremonial displays. Its location is at the heart of the whole park, centered along a sight line from the windows at the rear of the palace. Architect Giulio Parigi built it of a masonry foundation between

FIGURE 2–14 The original garden was laid out along a central axis symmetrically aligned with the Palazzo Pitti.

1630 and 1634. The large arena is surrounded by six tiered stands and a balustrade interrupted by twenty-four niches that originally held statues. In place of the existing stone steps for seating, the original amphitheater had ascending terraces planted with various tree species, including oak, ash, olive, and plane trees, which provided shade to the spectators.

Beyond the amphitheater, in line with the north-south axis, is the Forcone Basin. This basin, originally a fish breeding pond and collection basin for the waters provided by a nearby aqueduct, holds a statue of Neptune, god of the sea, by Stoldo Lorenzi da Settignano, circa 1565. Neptune, is depicted standing above a cliff with his trident raised, ready to pierce the earth to initiate a gush of water. Water does, in fact, originate from this sculptural piece, flowing down and out to the garden.

FIGURE 2–15 La Dovizia, or the Statue of Abundance, was added by Ferdinando II (1610–1670) to celebrate Tuscany's good fortune during his reign.

The Statue of Abundance (La Dovizia) (Figure 2–15) stands at the terminus of the visual line drawn from Pitti Palace along the length of the axis formed by the amphitheater and the Forcone Basin. Cosimo's son, Grand Duke Ferdinando I (1549–1609), commissioned the statue, which was begun by Giambologna in 1600 and completed by Pietro Tacca (1577–1640) and his assistant, Sebastiano Salvini da Settignano, years later. It was originally designed to reflect the image of Giovanna of Austria, wife of Francesco I and the mother of Marie dei Medici, but it was revised before being placed in the garden in 1636 for Ferdinando II. In its revised state, the sculpture depicts a woman standing proudly with a cornucopia of fruit and flowers and a sheaf of wheat. It was intended to represent the prosperity that Tuscany enjoyed during the reign of Ferdinando II, when the rest of Europe was suffering from war, sickness, and poverty.

The first stage of the garden's history ended under Ferdinand I, who was grand duke from 1587 to 1609. In the seventeenth century, Cosimo II (1590–1621), grand duke (1609–1621), commissioned the Giardino del Cavaliere (the Knight's Garden) near the city wall, forming the southern boundary of the garden. This building was used by Cardinal Leopoldo dei Medici and then by Cosimo III (1642–1723) as a place to meet with scholars, artists, and scientists. Just beyond the amphitheater, a path, oriented east to west, crosses the main axis. The eastern path terminates at the Kaffeehaus. This building, a later eighteenth-century addition to the garden, was designed by Zanobi del Rosso (1724–1798) as a place for the court to rest during walks in the garden.

From 1620 to 1640, under the direction of Cosimo II, Alfonso Parigi extended the garden westward beyond its previous boundary, the city wall of 1544. Parigi laid out a second axis at right angles to Tribolo's previous one, in the form of a long, sloping avenue known as the Viottolone, or Cypress Lane, descending to the Porta Romana (Roman Gate). This extended axis, flanked on either side by double rows of cypress and laurel trees, featured numerous ancient and modern sculptures regularly situated along its shaded path.

At the lower end of the avenue, Parigi created a large, oval pool that functions as a climax along the extending path. The pool is surrounded with a marble balustrade. Within this basin, he built the *isolotto* (little island) at the center, which is approached by two small bridges connecting it to the

FIGURE 2–16 The Fountain of the Ocean by Giambologna stands on the isolotto, *or "small island," behind the sculpted figure of Andromeda (foreground).*

FIGURE 2–17 The Moustache Fountain, sculpted by Romolo Ferucci del Tadda, was built as a series of drinking troughs for birds.

Viottolone. On the island stands the Fountain of the Ocean (Oceanus), father of the rivers of the world and ruler of the seas, by Giambologna (1529–1608). It was created in 1576 and placed there in 1637 with the statue of Neptune depicted rising above his pedestal, around which are arranged three marble river gods symbolizing the Ganges, the Euphrates, and the Nile rivers (Figure 2–16). Here water is again presented as a principal element in the garden. Sculptural god figures reinforce the power of Medici rule and their accomplishments in bringing water to the city of Florence. The garden of the isolotto marks the end of the avenue's vista. The area surrounding the pool is adorned with several hundred potted citrus trees, consistent with the original design.

On the upper end of the Viottolone, a crosswise path connects other significant garden features. The path, the Cerchiata Grande (Large Lattice Work), is named for the natural arbor of Holm oaks created along its length. This shady path links the wooded hunting area known as the "Ragnaia della Pace" and the Lemon House.

As one walks along the length of the Cerchiata, one encounters what is left of the Ragnaia della Pace. This area was once stocked with game birds that were hunted for sport with large nets. An interesting feature of the Ragnaia was the drinking troughs, the "Fontana de la Mostaccini" (moustache fountain), created for birds (Figure 2–17). These elegantly sculpted water troughs were built between 1619 and 1621 by Romolo Ferrucci del Tadda (d. 1621), who is responsible for many of the animal sculptures found throughout the gardens.

On the north end of the Cerchiata Grande is the Lemon House. Built from 1777 to 1778 after the design of Zanopi del Rosso, it is a rococo-style building that was used to store potted citrus trees during the winter months. The building, which retains much of its original appearance, is still used today for overwintering potted trees.

After, Gian Gastone, the second son of Cosimo III and the last of the Medici males, died in 1737, the Archduchy of Tuscany passed to the Habsburg-Lorraine. The Grand Duchy of Tuscany was given to Francis Stephen (1708–1765), Duke of Lorrain, who would later become the Holy Roman Emperor Francis I (1745–1765). During the Lorrain regency, the gardens were neglected. It was not until the reign of Leopoldo II that the gardens were restored and replanted, and new buildings were constructed. In 1775, Zanobi del Rosso built the Kaffeehaus in the northeast part of the gardens.

The nineteenth century brought about significant changes to the gardens. The French, who then controlled Tuscany, attempted to modernize Boboli by incorporating some of the popular English principles of garden design, which encouraged a more natural appearance. In the end, most of the gardens remained intact, with the exception of the wooded groves. In these areas, new tree species were introduced, and meandering paths were carved among them. Little has changed in the gardens since then, aside from the periodic restoration of specific areas by the Italian state, which now owns and maintains the gardens.

Chapter 3

Landscape Expression: The Renaissance Gardens of Rome

Figure 3–2 Three large rectangular water basins, called the "Fishponds," are situated along the first traverse axis at the base of the gardens.

breeding reservoirs for fish that were harvested for the cardinal's table (Figure 3–2). On the northeast end of this cross axis, sitting atop a tall hill, is the Fountain of the Organ, which has been recently restored (Figure 3–3). It is named for the music that it once made, which imitated the sounds of an organ. Designed by Ligorio, it is a baroque-style construction enhanced by a highly decorated façade, very original for sixteenth-century design. Multicolored panels depict mythological scenes with Orpheus, Marsius, Apollo, and others. At the center of the composition, artist and architect Gianlorenzo Bernini (1598–1680) designed an opening to house the hydraulic organ. The organ functioned through an elaborate system of conduits, toothed copper cylinders, organ pipes, and keys, all engineered by hydrologist Claude Venard. Beyond this first cross axis, the primary or central axis rises up the steep slope to the villa. Interrupting this line is the Fountain of the Dragons. It stands at the heart of the garden, an oval basin dominated by a fascinating sculptural group of dragons and surrounded by two semicircular flights of steps. The dragons were originally planned as an iconographical reference to the many-headed dragon conquered by Hercules. They acquired a

Figure 3–3 The Fountain of the Organ is named for the music that it once made, which imitated the sound of an organ. Below it is the Fountain of Neptune, added in the 1920s.

new meaning when finally constructed in 1572, prior to an anticipated visit by Pope Gregory XIII, whose coat of arms included dragons.

Just below the Fountain of the Organ is the Fountain of Neptune (Figure 3–3). Atilio Rossi, a respected curator for Villa d'Este, added this fountain in the 1920s. This creation succeeds in combining a marvelous water display with Ligorio's original waterfall connected to the hydraulic organ. Below the balustrade that stands along the terrace situated in front of the organ are three connected **nymphaeums** (richly decorated grotto gardens identifying with a nymph or spirit of nature that, according to ancient mythology, inhabited places of beauty, such as a garden or forest). At the base of the nymphaeums is a large water basin containing fountain jets from which water ascends to various heights. The water then falls into a succession of basins, with the torrent coming to rest and calmly flowing into the fishponds.

Just above the Fountain of the Dragons is the second major cross axis formed by the Avenue of the Hundred Fountains. One hundred jets of water feed three parallel channels said to represent the three rivers that merge into the Aniene, which runs from Tivoli to Rome (Figure 3–4). The Hundred Fountains border a long, straight path that extends from the Rometta Fountain to the Ovato, or Oval Fountain. Sculptures of obelisks, boats, and eagles, now covered with moss and ferns, represented significant symbols to the cardinal. The obelisks signify papal power; the boats, St. Peter; and the eagle, the Este coat of arms.

FIGURE 3–4 The Avenue of 100 Fountains is a major cross-axis in the garden.

FIGURE 3–5 The Fountain of the Rometta, or "little Rome," at the southern end of the first cross-axis. In the foreground is a scale model of the Tibertina Island in the shape of an ancient Roman boat.

On the northeast end of the Avenue of Fountains is the Ovato. This fountain, once called the "Fountain of Tivoli," also was designed by Ligorio. Its earlier name derives from its composition, which represents the mountains and rivers that make up the local landscape. Above the central waterfall is a statue of the Sibyl of Tiber (Sibyl Albunea) holding the hand of her son, Melicarte, who represents Tivoli. A half-moon terrace with a marble balustrade borders this rock construction. Water falls from the top of the fountain in a perfect domelike fashion. In the grottoes below are statues of reclining river gods that personify the major rivers of the Tiburtine region, the Aniene and the Erculaneo.

At the other end of the Avenue of Fountains, to the southwest, is the Fountain of the Rometta, or little Rome. This sculptural group, which only partly survives today, was built as a sculptural recreation of several significant buildings of ancient Rome, a period still revered by Italian culture. It was designed by Ligorio and constructed in 1570 by fountain maker Luzio Maccarone. Foregrounding the miniature city is a stream of water, which represents the Tiber, and a scale model of the Isola Tiburtina (Tibertina Island) in the shape of an ancient boat (Figure 3–5).

Along the principal axis, against the wall near the original entrance, is the Statue of Nature, the many-breasted Diane of Ephesus, which originally was situated at the Fountain of the Organ. The significance of this statue as a part of the water organ was likely an implication that water, the life-giving fluid of nature, was responsible for the sounds of the organ music.

After Cardinal Ippolito d'Este's death in 1572, the villa passed into the hands of his nephew, Cardinal Luigi d'Este, and then in 1605 to Cardinal Allesandro d'Este. Allesandro initiated a much-needed renovation as well as some new additions, such as the Biccherione Fountain, designed by Bernini. During the eighteenth century, the villa was sorely neglected. It passed to a German priest, Cardinal Gustov von Hohenlohe, to whom it remained until 1918. The property became a possession of the Italian state shortly after World War I.

Villa d'Este, today one of the best known of Italian villas, is an undisputed success in its contribution to place, culture, and the art of landscape design. It was also significant to the self-esteem of a very talented and visionary man. Ippolito d'Este's dream of a papal position was never realized, but his desire to create an estate grander than any other of its day was achieved to the fullest extent.

Villa Lante

1. Fountain of the Deluge
2. Houses of the Muses
3. Fountain of the Dolphins
4. Water Chain
5. Fountain of the Giants
6. Cardinal's Table
7. Fountain of the Lights
8. Palazzina Gambara
9. Palazzina Montalto
10. Parterre
11. Fountain of the Moors

Villa Lante
BAGNAIA, ITALY

Villa Lante at Bagnaia was built by the wealthy Cardinal Gianfrancesco Gambara in 1568, shortly after he became bishop of Viterbo. The lavish villa garden created by Gambara captures the spirit of mannerism, a new form of expression in the arts that developed in Renaissance Italy during the sixteenth century. This new "manner" of expression celebrated individuals and their accomplishments, rather than those of an entire society. At Lante, the gardens are a reflection of the creativeness, power, and prestige of Cardinal Gambara, whose wealth exceeded that of most other cardinals of his era. He spared no cost in developing a pleasurable retreat. Letters written by Cardinal Gambara document his ambitious plans for a residence, garden, and park that would be beyond comparison to any similar projects.

Villa Lante is located in Bagnaia, Italy, a small town near the city of Viterbo, 50 miles from Rome. During the sixteenth century, the town was controlled by the bishopric of Viterbo. It developed into a popular summer destination for many of the bishops, who made it their choice for a country retreat. By 1514, a growing interest of these religious leaders to entertain their visitors and the rising popularity of hunting for sport led to the development of a hunting park. In 1514, Cardinal Raffaele Riario had 25 hectares (62 acres) on the nearby hillside of Monte Sant' Angelo, just south of town, enclosed with walls and filled with wild game. A small and simple lodge was built to serve invited guests on the occasion of a hunt.

Further development of the park occurred under the direction of Cardinal Niccolo Ridolfi (d. 1550), who became bishop of Viterbo in 1549. Ridolfi was responsible for bringing water to the park as well as to the town of Bagnaia by commissioning the construction of an aqueduct to carry water from nearby springs. However, it was Cardinal Gianfrancesco Gambara (d. 1587) who, upon receiving the bishopric of Viterbo, set out to create a villa and gardens within the larger wooded hunting park. The design of the first of a pair of twin ***casini*** (the principal dwelling in an Italian villa garden) and the associated formal garden that makes up the villa complex is credited to architect Jacopo Barozzi, called "Vignola" (1507–1573). Though no specific documentation has been found, there is sufficient evidence to justify Vignola's involvement. While he was responsible for the design, an architect and a hydraulics engineer from Siena, Tommaso Ghinucci, is believed to have directed the work, which began the year of Vignola's death, 1573. The construction of the formal garden was nearly completed by 1578, the same year Pope Gregory XIII visited the Bagnaia estate. Only one of the proposed

Villa Lante Time Line

1514	Cardinal Raffaele Riario creates a 62-acre hunting park on the hillside of Monte Sant' Angelo.
1549	Cardinal Niccolo Ridolfi (bishop of Viterbo in 1549) brings water to the park and to the town of Bagnaia by having an aqueduct constructed to carry water from nearby springs.
1568	Cardinal Gianfrancesco Gambara begins construction on the villa and garden within the larger wooded hunting park at Bagnaia. Architect Jacopo Barozzi (called "Vignola") and architect and hydraulics engineer Tommaso Ghinucci are employed to design the house and gardens.
1578	The formal garden is completed.
1587	Cardinal Gambara dies, and the pope assumes control of the estate. Occupancy is granted to Cardinal Federic Comaro.
1590	Cardinal Allesandro Paretti Montalto, nephew of Pope Sixtus V, takes up residence at the villa. Montalto completes the second of the twin *palazzine* that bears his name.
1656	The villa passes to the Lante della Rovere family, from which it derives its current name.
1953	The villa is purchased and later restored by the Societa Villa Lante.
1973	The villa becomes a possession of the Environment and Architecture Property Service of the Region of Lazio, which continues to attend to the restoration and conservation of the estate. The gardens open to the public.

pallazine (small palaces) was completed, Palazzina Gambara. Documents from the papal visit indicate that a second casino bearing a likeness to the first was proposed, but it would not be realized until Cardinal Montalto took up residence at the villa in 1590.

Upon the death of Gambara in 1587, the pope assumed direct control of the estate, and its occupancy was granted first to Cardinal Federico Cornaro until his death in 1590 and then to Cardinal Allesandro Paretti Montalto, the nephew of Pope Sixtus V. The papal authority expected residents of the estate to both contribute to and maintain the villa, respecting the original intent of its development. Montalto therefore made some changes in the garden, particularly the repair and alteration of several fountains and the completion of the second of the twin *palazzine* that bears his name.

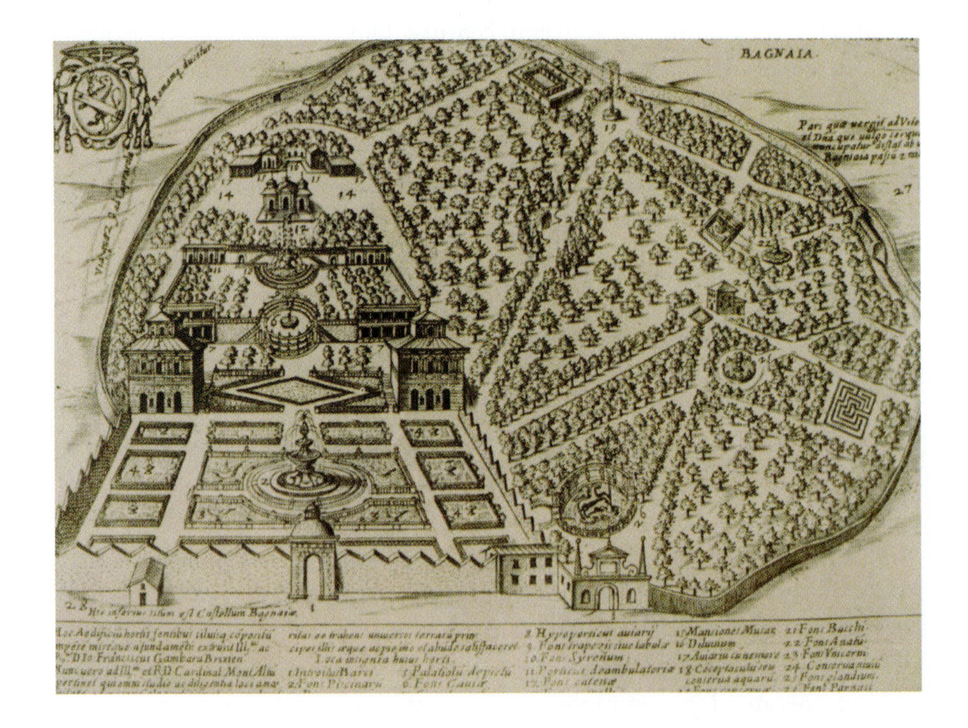

Figure 3–6 Giacomo Lauro, Villa Lante, Bagnaia, from Antiquae urbis splendor, *Rome, 1612–1628. Courtesy of Dumbarton Oaks, Research Library and Collection, Washington, DC.*

From a design standpoint, Villa Lante presents an interesting juxtaposition of the strictly ordered nature in the formal garden with a more natural style in the hunting park. The two compositions, existing essentially side by side, represent the contrasting opinions of garden design that would become the subject of great debate in Europe in centuries to come (Lazzaro 1990) (Figure 3–6). Two distinctly different views toward nature are revealed, that which embraces a harmony with "wild" nature and that which imposes artistic intentions on nature to satisfy the human desire for creativity. The park embraces nature attempting to recreate its informal character. It is peaceful and mostly unadorned. In the formal garden, nature is controlled. Renaissance order and balance are successfully achieved along a uniquely designed central axis, unique in that water has been used to create this spine. Water is the organizing element throughout the garden, linking the various garden compositions as it appears throughout the garden in varied forms.

Figure 3–7 The Fountain of the Giants represents the Tiber and Arno rivers, symbolizing the friendship between Rome (the papacy) and Florence (the Medici family).

The design of the garden is faithful to the Renaissance principles of order and geometry, with a distinguishable influence of ancient classical themes. Ancient models, particularly the classic forms and symbolic associations encountered in classical work, appear to have inspired parts of Vignola's design.

The formal hillside garden is laid out as a series of symmetrically organized terraces. Each is composed of unique experiences skillfully connected by water. The water theme originates at the Fountain of the Flood (also called the "Grotto of the Deluge") on the uppermost terrace of the garden and continues down the central axis to the garden's other levels. From here the water flows to the Fountain of the Dolphins, so named for eight pairs of dolphins cast along the base level of this multitiered fountain.

The water reappears along the center line of a ramped stairway that descends the hillside to the next garden level. At the center of the stairs a raised stone channel bordered on either side by a series of elegantly carved volutes carries the water down the hill. At the top of the channel the water flows from the head of a crayfish, or *gambero,* as it is called in Italian. This is a creative gesture that makes reference to the cardinal's family name. This splendid garden feature is known as the Catena di Aqua (Italian for "water chain"). It solicits water as a medium for the creation of a beautiful work of art. Water is essentially molded by the shape of the channel's border, inducing a swirl-like movement that reflects the sunlight like a precious jewel. It is indeed a unique creation.

The water is made visible again on the terrace of the Great River Gods—the monumental Fountain of the Giants. Water streams down from the water chain, splashing over stepped basins and coming to rest in a large semicircular basin flanked by two massive sculptural figures reclining against a tall stone retaining wall. These giant figures represent the Tiber and Arno rivers, symbolizing the friendship between Rome as the center of Christian faith and Florence as Italy's capital of industry and wealth (Figures 3–7, 3–8). A double staircase, perpendicular to the main axis, is located just behind the stone wall to provide access to this garden level.

Figure 3–8 The Fountain of the Giants.

FIGURE 3–9 The Cardinal's Table, designed to accommodate outdoor dining.

Centered on the garden axis, just beyond the Fountain of the Giants, is the Cardinal's Table, a rectangular stone table with a long water basin carved into its top (Figure 3–9). Outdoor dining, popular among ancient Romans, continued as a tradition during the Renaissance. Many gardens were designed to accommodate this desire for *al fresco* (fresh air or outdoor) dining. At Lante, Church dignitaries gathered around the so-called Cardinal's Table on sunny summer afternoons to enjoy fruits and wine chilled by the crisp, cool waters that filled the channel.

Joining the garden level to the casino level below is the Fountain of the Lights. The fountain, located on the central axis, is set into the hillside of the terrace, upon which the Cardinal's Table is situated. It is constructed as a series of concentric tiers, built into and out from the hillside in concave and convex manners, respectively. The many delicate water jets rising from the stacked tiers resemble ancient oil lamps. It is this distinction that gives the fountain its name. The twin *casini,* Palazzo Gambara and Palazzo Montalto, occupy two levels within the garden. The Fountain of the Lights descends upon the uppermost level, called the "piazza" level. The lower level is reached by descending a grassed slope that lies between the twin buildings.

A hedge-lined path traverses the hillside to the lowest garden terrace situated at the base of the slope. Sixteen equal garden sections composed of elegant and meticulously groomed *parterres* (Figure 3–10) are organized around a monumental fountain called the "Fountain of the Moors." Four male figures called the "Moors" proudly display Cardinal Montalto's family symbol: lions, mountains, and pears crowned by a star. Water spurts from the points of the star, falling gently and elegantly into the basin below. This lower garden terrace, now the final area of the garden to be experienced by visitors today, was the very first to be seen by the cardinal's guests who entered the garden from the city center, which lies just on the other side of the walled enclosure.

In 1656, the villa, occupied by papal appointees until the mid-seventeenth century, passed to the Lante della Rovere family, from which it derives its current name. It remained their possession until 1953, when it was purchased and later restored by the Societa Villa Lante. In 1973, it became a possession of the Italian state. The formal villa garden is the sole destination for nearly all visitors today, as the park has become mostly overgrown and neglected.

FIGURE 3–10 The parterres are composed of evergreens clipped into ornate patterns.

Chapter 4 Landscape Expression: Gardens of the Italian Baroque

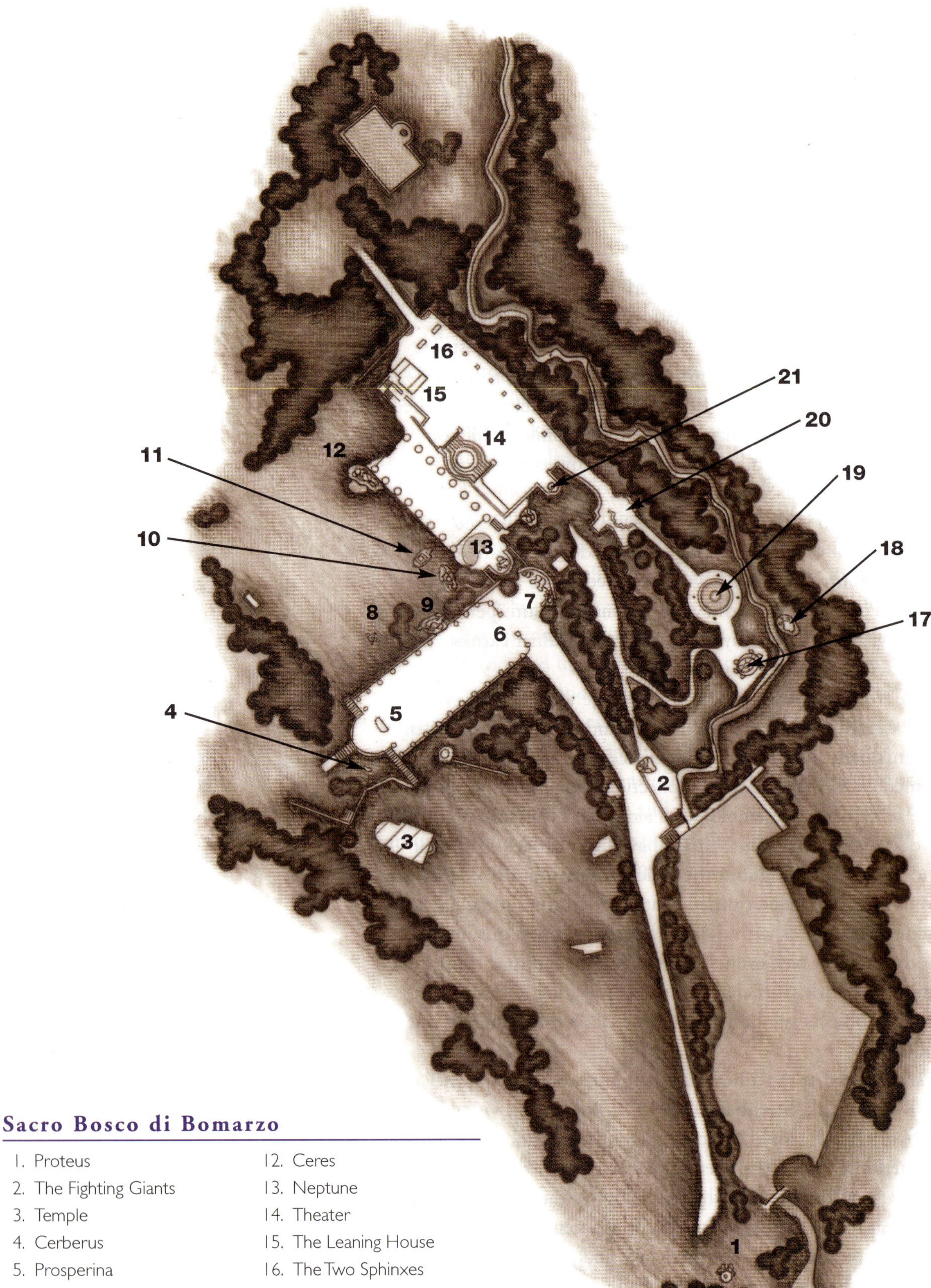

Sacro Bosco di Bomarzo

1. Proteus
2. The Fighting Giants
3. Temple
4. Cerberus
5. Prosperina
6. Heraldic Bears
7. Echidna-Lions-Fury
8. Cantero
9. Orc
10. The Fighting Dragon
11. Elephant with Warrior
12. Ceres
13. Neptune
14. Theater
15. The Leaning House
16. The Two Sphinxes
17. Giant Tortoise
18. Muzzle of an Orc
19. Pegasus
20. The Three Graces
21. The Sleeping Nympth

dedicated to Giulia, Orsini's beloved wife, who died in 1562, 20 years after work on the garden began. Battling giants, love goddesses, and temples—these various expressions encourage our contemplation of such powerful forces of life.

References also are made to literature for clarification of the meaning of these weird sculptural allusions, associating them with pieces from Aristo's epic poem *Orlando Furioso* (1532), Rabelais's *Gargantua* (1534) and *Pantagruel* (1532), and the *Hypnerotomachia Poliphili* by Francesco Colonna in 1499. For example, the sculpture of the warrior brutalizing a young man could very well be related to a scene from *Orlando Furioso,* where Orlando, gone mad, slaughters another to ease his heart's pain.

The literary symbolism in the park, familiar to intellectual circles in Orsini's time, is diluted or perhaps even lost for today's visitor, who knows the garden as Parco dei Mostri (monster park), a fanciful collection of magical-looking beasts engaged in strange activity. Scholars have acutely criticized this simple exploitation of an intellectual masterpiece. It is true that the garden's sculptural symbolisms likely allude to a deeper meaning, but we must not only regard this work as an intellectual experience or as a statement or an answer to something. This marvelous work is not only about something; it *is* something, something quite fascinating. Its composition denied all rules in its making. Perhaps we can recapture that tradition and free ourselves from the need to explain or justify its creation. Its interpretation is unclear, but its accomplishments are obvious. Sacro Bosco is an artistic wonder, unprecedented and unrivaled to this day.

Progression through the park requires us to select our path with no clear itinerary except that provided by its current owner, Giovanni Bettina. Upon entering, one passes through a gate decorated with the crest of the Orsini family. Two sphinxes are situated beyond the gates to both greet and bid farewell to visitors as they leave the park. Immediately to the left (south) is a colossal head sculpted from rock. The head, believed to represent Proteus, the son of Neptune and servant of Poseidon, is sculpted with its mouth wide open and its eyebrows, cheeks, and lips all ending in swirling waves (Figure 4–1). Here Proteus displays his incredible strength, balancing the world upon his shoulders. On top of the earthlike globe is a castle, thought to represent Orsini's palace and thus displaying a symbol of the power of the Orsini family in the world. To the north is the Fighting Giant (Figure 4–2). This colossal figure depicts a warrior brutalizing a young man. Perhaps this powerful figure has gone mad, unaware of his actions, or perhaps he knows very well what he is doing, defeating those who perpetuate evil.

FIGURE 4–1 *Proteus balances the world atop his head as a symbol of the power of the Orsini family.*

The garden path continues northeast, arriving at a collection of sculptures made to represent a tortoise and a whale. The tortoise faces a ravine, from which a whale rises, eager to devour it. It is likely the tortoise, protected by its thick shell, is meant to remind us of the brave Roman soldiers who protected themselves in battle with shields held in front of their heads. Not far from the tortoise is the winged horse, Pegasus. This is said to be a tribute to the Farneses, the family of Orsini's wife, Princess Giulia. Pegasus was the emblem of the Farneses's coat of arms. To the west is a monumental sculpture depicting Neptune, god of the sea. Here Neptune most likely represents the Tiber, the river of central Italy.

FIGURE 4–2 The Fighting Giants represent the ongoing battle between good and evil. Here evilness is defeated at the hands of the brave warrior.

The theater, a recreation of a classic Roman theater, is located at the north end of the garden. It is constructed as a series of concentric steps built into a convex-shaped retaining wall. This area of the site is believed to be the old entrance to the park. The Leaning House stands adjacent to the park's original entrance (Figure 4–3). It was intentionally built on an angle, appearing to have sunken into the earth. The forces of nature are humorously represented here. It is a reminder of the reality of life and one's place in the world. One is called to recognize that human accomplishments are not superior to nature.

To the south is a dragon sculpture. In ancient myths, the sleepless dragon was always regarded as brave and trustworthy. It often was included in Renaissance period gardens as a protector of the garden or of sacred ground. At Bomarzo, the dragon is depicted in battle with a lion, defending the sacred wood. Occupying the same space as the dragon is the elephant, historically associated with warfare. The giant sculpted figure is depicted here with a dead soldier in its trunk, a representation of victory in battle.

FIGURE 4–3 The forces of nature are shown to be more powerful than human accomplishments with the humorous construction of the Leaning House.

Near the dragon and elephant is the frightening Orc (Figure 4–4). This sculptural piece seems to signify the gateway to hell. The figure, with its mouth wide open, eagerly awaits visitors. Situated in the hollow darkness of its gaping mouth is a stone dining table. This, a luring device, causes one to ignore the frightful exterior in favor of an open invitation to feast and partake in worldly pleasure within the darkness. This seems to represent the worldly desires that often lure Christians away from their faith and into the darkness of hopeless eternity.

Other sculptures within the park include Echidna, a woman's figure, with snakelike appendages signifying lust and seduction; the sleeping nymph, depicted as seemingly one with the earth, an expression of nature in human form; Ceres, goddess of harvest, signifying the good fortune and gifts that come from the earth; the Cantero (giant vase), perhaps representing the journey of Bacchus, the god of merrymaking and wine, who carried a cup much like this one down to hell; and a sculptural representation of Cerberus, the three-headed guardian of hell. Further emphasizing the tranquility of nature enjoyed in the sacred woodland is the figure of Prosperina, revered as the enemy of evil. Here the goddess welcomes the visitor to a seat of peaceful repose.

The journey through Orsini's ingenious creation concludes at the temple that he built to honor the memory of his wife, Giulia Farnese. It is now also dedicated to the memory of the current owner's wife, Tina Severi Bettini, who died in 1987 following an accident that occurred while supervising the restoration of the garden.

FIGURE 4–4 The Frightening Orc lures visitors into the darkness of its gaping mouth, believed by some to represent the entrance to hell.

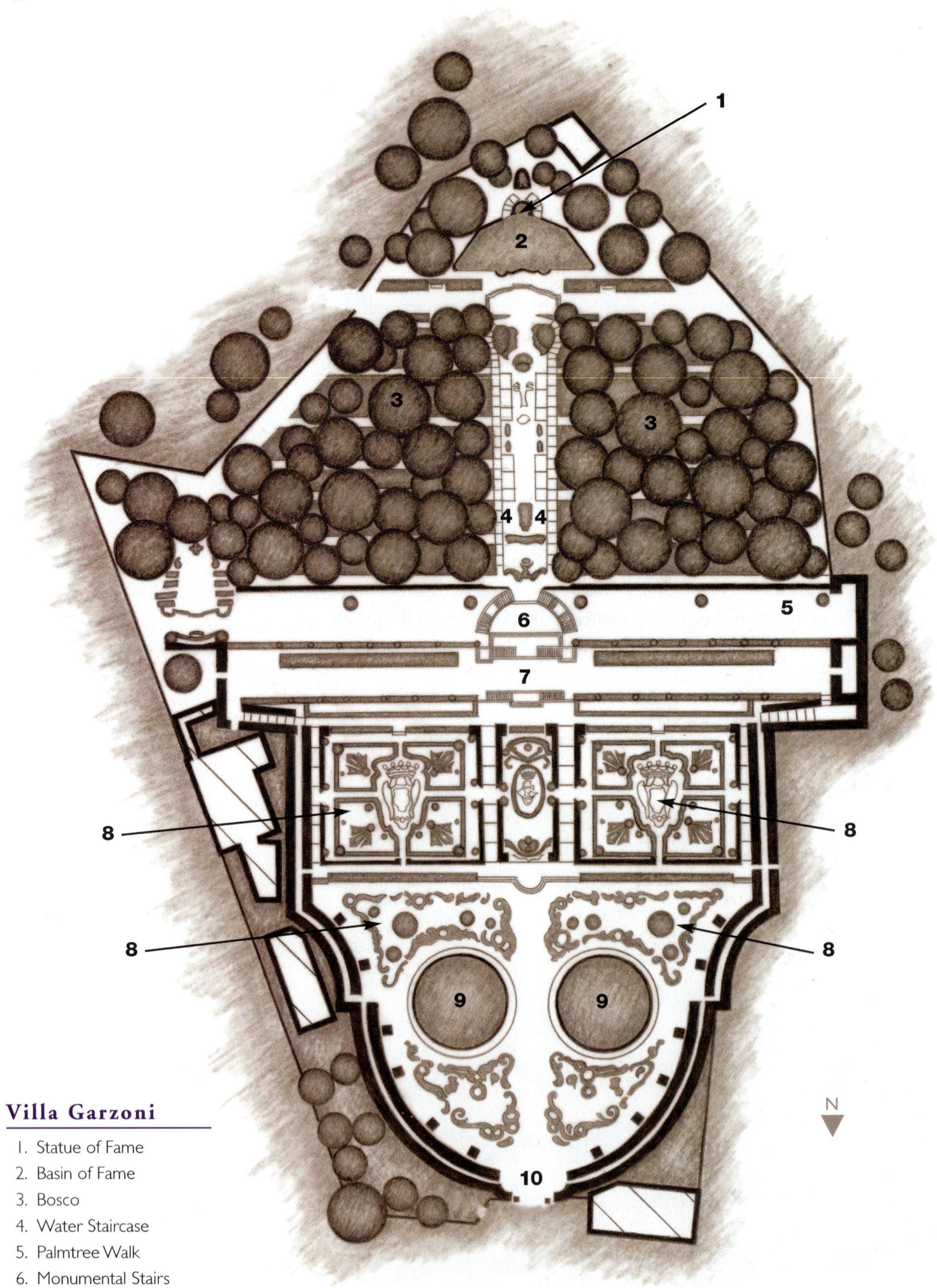

Villa Garzoni

1. Statue of Fame
2. Basin of Fame
3. Bosco
4. Water Staircase
5. Palmtree Walk
6. Monumental Stairs
7. Pomona Walk
8. Parterres
9. Water Basins
10. Garden Entrance

Villa Garzoni

Collodi, Italy

Lucca, as a growing trade center, was an economically secure state in the seventeenth century. A period of new and extravagant building conveyed its wealth, of which Villa Garzoni in Collodi, near Lucca, is an example.

The seventeenth-century garden of Villa Garzoni conveys a spirit of Renaissance with the emerging baroque style. The gardens, while generally faithful to Renaissance proportions and geometries, have a level of ornamentation that moves beyond that found in classical gardens of the previous centuries. The design is open and inviting with a baroque spirit that draws one into the space at first glance. Its extravagance and ornate qualities are immediately revealed in the colors and detail of constructed and planted elements that make up the garden.

The site plan is unusual, in that the garden seems to have been planned as a separate experience apart from and unrelated to the house (Figure 4–5). The house, built on the site of an ancient medieval castle, is located atop a rocky slope that would have made creating gardens near it a very difficult task. Little is known about the first phase of the design of this villa complex. An early sixteenth-century survey does indicate a modest garden associated with the house. A later survey, done in 1692, and a detailed plan of the property the same year show the garden as it exists today.

Villa Garzoni Time Line

1633	The Garzoni family begins making improvements to the existing house and gardens.
1692	A survey of 1692 and a detailed plan of the property made in the same year document the layout of the garden.
1730s	Architect Ottaviano Diodata is employed by the last Romano Garzoni to redesign the original garden, laid out in the mid-seventeenth century.
1945	The Garzoni property is sold.
1994	The garden is restored.

Figure 4–5 A period painting of the Villa Garzoni. Courtesy of Villa e Giardino Garzoni srl.

Figure 4–6 A fanciful parterre *at the garden's entrance displays the emerging baroque style of design.*

The design of the garden, which occupies a hillside apart from the house, takes advantage of the sloped terrain, producing a succession of terraces linked by a series of elaborate stairways. The central staircase is a major garden feature, exhibiting a baroque quality, emphasized by its elaborate ornamentation and details. The entrance to the garden is at its lowest terrace, designed as a colorful *parterre* dominated by two large circular pools on either side of the central path (Figure 4–6). Boxwood hedges defining the lower garden's boundaries are trimmed into a variety of whimsical forms, many of which represent animal figures. The garden rises uphill toward the majestic staircase. This section of the garden is decorated with the Garzoni family's coat of arms, created with colored pebbles and flowers. The triple flight of stairs ascends the hillside, joining the multiple terraces. The first terrace lined with palm trees is called the "Palm Tree Walk" or "Avenue of Palms," frequently praised for its exotic beauty. But even more significant is the form of these trees. Their outline, set apart from the background vegetation, makes a splendid contribution to the planted landscape. Moreover, the trees create a focus for one's eyes on the landscaped terraces in the foreground.

The second terrace, now in an overgrown state of neglect, was once decorated by numerous sculptures. It is called the "Pomona Walk." Pomona, who stood at one end, was regarded as the guardian goddess of the garden. The water staircase flanked by two stone sculptures personifying Florence and Lucca is the main feature of the final ascent. It is a continuation of the central axis that culminates with a massive statue of Fame, or Fama, the Roman personification of rumor, who brings water into the garden by blowing into a seashell (Figure 4–7).

Figure 4–7 The statue of Fame, perched high atop the hillside, brings water into the garden by blowing into a large seashell.

A winding path links the upper part of the garden to the house. The path crosses a beautiful pergola-type bridge made of brick and stone from which one may pause to enjoy framed views of the forest. A suspended bridge transports visitors over a ravine and into a dense bamboo forest. From here one may venture into an intricate maze before arriving at the house. The house, built on the site of an ancient castle, sets high on the hill, with a commanding view of the town of Collodi and the distant hills of the Val di Nievole (Figure 4–8). Ottaviano Diodata, an architect employed by the last Romano Garzoni (1721–1786), is associated with the final phase of work on the garden, which included a new hydraulic system to feed the fountains.

FIGURE 4–8 The house, built high atop the hill, has a commanding view of the town of Collodi.

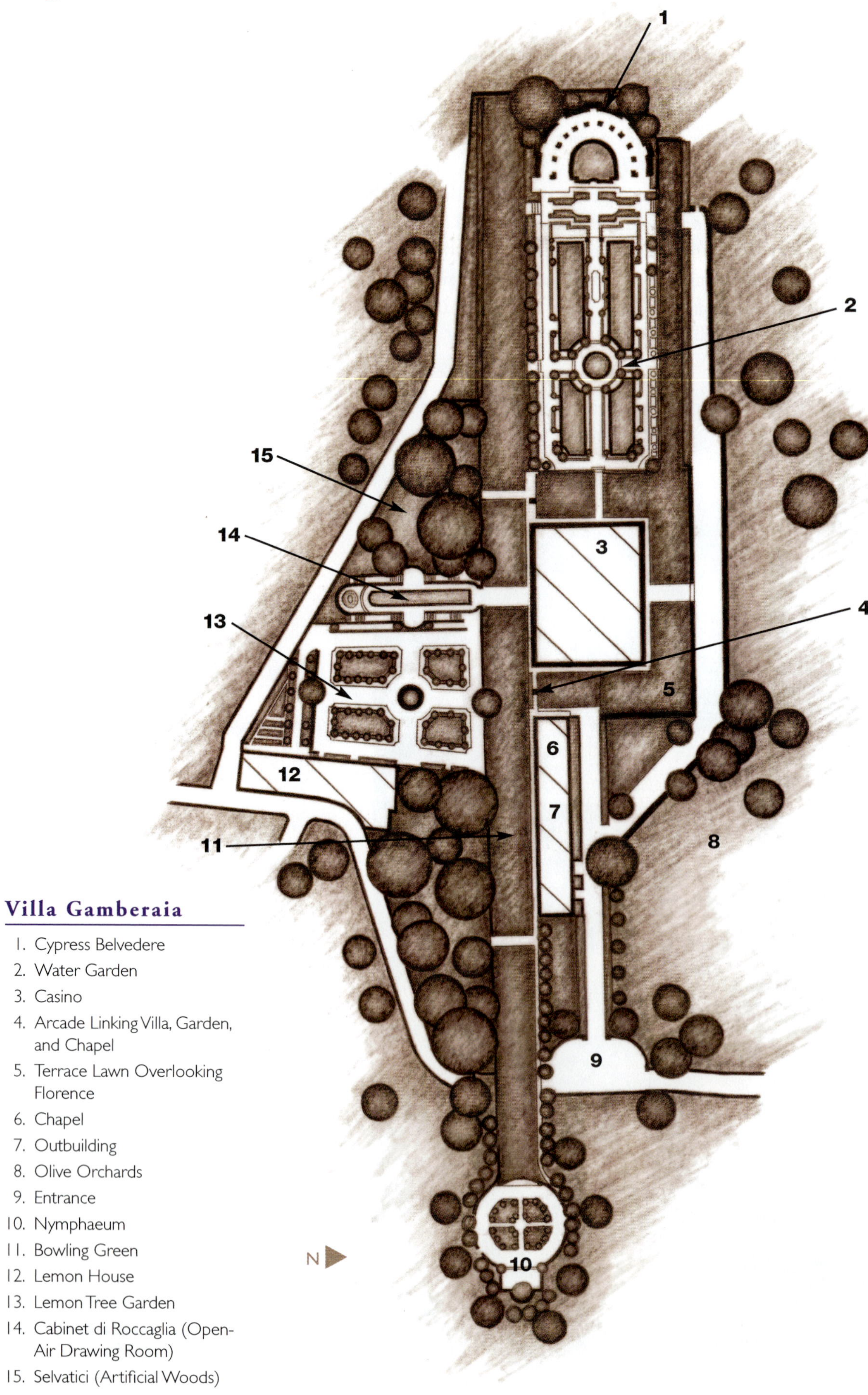
1
2
15
14
3
4
13
5
6
12
7
8
11
9
N
10
Villa Gamberaia
1. Cypress Belvedere
2. Water Garden
3. Casino
4. Arcade Linking Villa, Garden, and Chapel
5. Terrace Lawn Overlooking Florence
6. Chapel
7. Outbuilding
8. Olive Orchards
9. Entrance
10. Nymphaeum
11. Bowling Green
12. Lemon House
13. Lemon Tree Garden
14. Cabinet di Roccaglia (Open-Air Drawing Room)
15. Selvatici (Artificial Woods)

Villa Gamberaia

Settignano, Italy

Villa Gamberaia possesses almost every characteristic of excellence attributed to the Italian villa garden. It is a small domestic garden, yet its spaces are so splendidly designed and proportioned with a variety of designed effects: open, sunlit terraces, shady groves and green loggias, water features, and enchanting views of distant horizons.

Originally a fourteenth-century farmhouse owned by an order of Benedictine nuns, this small villa complex in Settignano changed hands several times before being purchased by Zanopi di Andrea Lapi in 1610. It was Lapi who had the first gardens laid out to complement the newly restored villa house. A century later, Vincenzio Maria and Piero Capponi became the owners of the villa. The Capponis made few changes to the house, but they did enlarge the garden. The property remained in the Capponi family until the middle of the nineteenth century, and it changed hands many times again before the end of that century. In 1896, Catherine Jeanne Keshko, wife of Prince Eugene Ghika, purchased the villa. Together with Martino Porcinai, her head gardener, Ghika created the estate's main garden on the south side of the house.

During World War II, a retreating German officer ordered the burning of the villa, which had been used to store German army charts and maps. All but the outside walls of the house were severely damaged, as well as the surrounding gardens. Marcello Marchi, a wealthy businessman, purchased the villa in 1954 and began an extensive restoration project that lasted over a year. Much of the original framework was still evident, enabling Marchi to retain most of the original design (Figure 4–9).

Villa Gamberaia Time Line

1610	Zanopi di Andrea Lapi purchases the property at Gamberaia from Francesco di Baccio di Manente Buondelmonti and lays out the first garden to complement his newly restored villa house.
1717–1725	Vincenzio Maria and Piero Capponi become the owners of the villa upon the death of Giovanni di Andrea Lapi and considerably enlarge the garden.
1896	Catherine Jeanne Keshko, wife of Prince Eugene Ghika, purchases the villa. Together with Martino Porcinai, her head gardener, Ghika creates the estate's main garden on the south side of the house.
1944	A retreating German officer orders the burning of the villa, which had been used to store German army charts and maps during World War II.
1952–1954	The Holy See aquires the villa.
1954	Marcello Marchi, a wealthy businessman, purchases the villa from the Holy See and begins an extensive restoration project.
1955	The restored garden opens to visitors.
1988	The villa is inherited by Marcello Marchi's heirs.
1994	The villa passes to Marcello's daughter, Franca, and her husband, Luigi Zalum, who continue the work of restoration and conservation.

Figure 4–9 Engraving by Giuseppe Zocchi (1744) depicts a view of the villa from the Via del Rossellino. Courtesy of Villa Gamberaia.

FIGURE 4–10 The Cabinet di Roccaglia extends the indoor living space to the outdoors.

The villa today characterizes the simple beauty of the classic Tuscan style. The house, quite simple in appearance, is a block-like construction, standing on walled foundations. These walls retain the earth that forms the grassed terraces adjacent to the house. The west façade opens to marvelous views of the Tuscan countryside. Directly through the main entry an inner courtyard leads to another door that provides entry into the garden. Here the house is essentially extended into the garden by the creation of an intimate garden room called the "Cabinet di Roccaglia," added by the Capponi family in the 1700s (Figure 4–10). This sunken garden links the ground level to the upper garden. Niches containing terra-cotta statues interrupt its curved, grotto-like walls of colored pebbles and shells. On either side are stairways that provide access to an upper level. To the south is a dense grove of trees. To the north is the lemon garden, where lemon and orange trees are set out in terra-cotta pots during the summer months. These pots are moved to the nearby Lemon House during the winter to shelter them from the cold.

From the first floor of the villa's south façade, two balconies extend over an arcade supported by piers, with views out to the neighboring countryside. The arcade joins the first floor to the main garden by a spiral staircase enclosed by the external south pier. This main garden, often called the "water garden," was created by Princess Ghika to replace an eighteenth-century *parterre de broiderie* garden (Figure 4–11). In this south garden clipped boxwoods border the garden paths that separate the rectangular space into four elongated pools, the center of which is marked by a small rustic fountain. These pools hold still water, inconsistent with the character of most water features found in Italian gardens. Yet here in this "modern garden," as it was

FIGURE 4–11 The main garden created by Princess Ghika during the late nineteenth century.

FIGURE 4–12 A view north from the green arcade toward the house.

labeled in a plan created in 1918, it is the stillness of these water features that contributes to the garden's peaceful charm. The garden is accented by **topiary** (trees or shrubs tightly clipped into a variety of shapes that appear almost sculpture-like) work and border plantings of lavender, iris, lilies, roses, and oleanders, some planted in pots (Figure 4–12). On the south end of the garden is a semicircular water lily pond, behind which stands a cypress belvedere of the same shape. This neatly clipped green arcade frames the garden and provides a shady respite from which a magnificent view to the south, over the Arno valley, can be enjoyed.

Parallel to the house, on its east side, is a long turf *viale* (avenue), itself an interesting garden space. It reaches from one end of the garden to the other. Piero Capponi also added the nymphaeum, a richly decorated grotto garden, in the 1700s. It marks the north end of the grassed avenue. The curved wall of the nymphaeum contains a niche in which a figure, traditionally identified as Neptune, is depicted brandishing his trident. At the southern end of the garden the path terminates in a balustrade with a view of the Arno valley.

The garden of Villa Gamberaia succeeds as a sequence of skillfully proportioned spaces, harmoniously linked to create a beautiful composition. The garden is revealed to us as a progression of charming outdoor rooms, designed to awaken our senses by introducing wonderful colors, spectacular views, and a perfect balance between openness and intimacy.

Chapter 5 Landscape Expression: Modern Variations

The Getty Center

1. Helicopter Pad
2. Service Road and Tramway
3. Arrival Plaza
4. Restaurant
5. Upper Garden and Grotto
6. Museum Courtyard with Fountain
7. Gardens
8. South Garden
9. South Viewpoint
10. Garden Terrace Café
11. Irwin's Central Garden

The Getty Center

LOS ANGELES, CALIFORNIA

The Getty Center Art Complex development in Los Angeles, California, built during the 1990s, is a product of European influence that is both classic and modern in its theories and approach to contemporary design. The designers of the project identified with the same theories held by Renaissance humanists, who believed that society is enlightened, enriched, and inspired by extraordinary human accomplishments. The center's architect, Richard Meier, created a design for the complex based on his philosophy that museum architecture should "encourage the discovery of aesthetic and cultural values" (Brawne 2000, 9). Meier's buildings enhance a visitor's experience of displayed art by creating spaces that are artful in themselves, strongly contributing to, but not overwhelming a museum's collection. The Getty is unique in that the building and the site are integrated as one art form, recalling the Renaissance spirit of design. The overall plan of the complex encourages human interaction, and it both facilitates and contributes (as a work of art itself) to the Getty mission of "serving the public in the understanding, enjoyment, and preservation of the world's artistic and cultural heritage" (Brawne 2000, 9).

FIGURE 5–1 *The Italian Travertine stone provides a connection to the landscape. Courtesy of Getty Communications.*

The Getty complex is informed by the natural context of its site. It occupies a hilltop in the foothills of the Santa Monica Mountains, overlooking the Los Angeles coastal basin and the Pacific Ocean. The formal organization of the center is dictated by the site's topography. The six buildings of the complex are aligned on axes that coincide with two natural ridges—one lining up with the street grid of Los Angeles, and the other running parallel to the San Diego Freeway, just below the site. The buildings are undeniably influenced by Mediterranean tradition, with clean, straightforward architectural forms and warm, natural colors complemented by the sunlight. The exterior is clad with a Travertine stone from Italy that gives the building a sense of permanence and stability in the landscape (Figure 5–1). Complementing the traditional stone exterior are off-white painted metal panels and large panels of glass, modern variations identifying with the architectural style of contemporary California.

From the outset, the spaces between the buildings became vital to the overall character of the site. Gardens and courtyards were designed to harmonize with the various parts of the complex and to establish areas of human scale (Figure 5–2). Like the buildings, the gardens seem inspired by Mediterranean tradition, especially the achievements of the Italian Renaissance villa gardens. In fact, the designers of the Getty complex traveled to Italy early on in the design process to explore several of the great gardens, including Villa d'Este in Tivoli and Villa Lante in Bagnaia. These hillside estates provided inspiration and historic precedents in a setting similar to the

FIGURE 5–2 *Gardens and courtyards establish a sense of human scale adjacent to the building. Courtesy of Getty Communications.*

Figure 5–5 A vine-covered pergola provides both shade and a sense of enclosure to the space. Courtesy of Getty Communications.

white or mauve to complement the flower colors of the vines that trail over them (Figure 5–5). The museum courtyard is a California equivalent of the splendid piazzas throughout Italy. People sit at small tables shaded by white umbrellas and cooled by the site and sound of water from an adjacent pool lined with spouting jets.

While the designed site is a product of a classical approach to design, there is still a unique mix of modern ideas with clear expressions of local context. The desert cactus garden on the south promontory of the center, for example, conveys a recognition of and appreciation for the natural composition of the local landscape. It is an idealized version with cactus and a variety of succulents skillfully laid out to represent a desert landscape.

The building complex and its associated landscape culminate in a rather unusual but intriguing and provocative way. The design for the central garden, originally intended as the space that would unite the entire concept envisioned by Meier, was awarded to California artist Robert Irwin. Irwin developed a garden that clearly opposes the rational order and strict geometries that define the rest of the Getty complex. His design challenges the permanence and stability established throughout the site. In fact, Irwin's garden celebrates change and introduces diverse experiences that reveal the passage of time and evidence of transition (Figure 5–6). For example, the trees in his garden are deciduous and thus provide clear evidence of the passage of time through the activities of seasonal growth. The activities of renewal are celebrated in spring with a spectacular collection of blooms and leaves in colors that Irwin refers to as the ***"pièces de résistance"*** (an outstanding accomplishment). Irwin labels his work "a sculpture in the form of a garden aspiring to be art," yet it seems here that it is art that has sought new expression. It is indeed a work of art, (to use Irwin's own words) something "that is not limited to a particular instance . . . [something that is] cumulative over a lifetime (Weschler 2002, 8–9)." Irwin has found in nature a new expression of art that is beautifully dynamic, a work

FIGURE 5–6 Aerial view of Getty Center Central Garden. Courtesy of Marshall LaPlante, © 2003, J. Paul Getty Trust. Garden © Robert Irwin.

that is never completed or truly controlled. It is an intentional contrast to the rest of the development, where beauty is embodied in the essence of refinement and the product of human innovation and control.

As a result of its hilltop setting, the Getty Center conveys an almost immediate sense of importance and privilege or prestige. Yet despite its location, the Getty does not set itself apart from the rest of the community. In fact, its location actually provides an extraordinary connection to the city. Not only is it easily accessible from the neighboring cities, but they are also the subject of its views. The center has established itself as an integral part of local urban fabric and has succeeded in its mission to develop a complex that is more than buildings that house art. Its main attraction of course is its art museum, which ranks among the finest in the world, but its gardens, plazas, and hillside prospect offer many more opportunities that strengthen its connection to the people of Los Angeles and around the world.

Chapter 6

The Urban Environment: Piazzas of Italy

Evolving since the Middle Ages, the piazza, or town square, has become an important part of many Italian cities. From the early days of the Romans, city planners were concerned with organizing space for public activities. Many functions, such as government events, religious ceremonies, and public markets, were typically staged outdoors. At the onset of the Middle Ages, a definite spatial pattern emerged within the city. Newly constructed buildings began to articulate areas of open space within the city fabric. Often these areas were intentionally preserved and functioned as a parvis for the new buildings. The piazza was, and still is, a place for people to gather, an organized space given life through public activity. Warm climatic conditions, coupled with the outgoing and very social personalities of the Italian people, likely contributed to the sustained popularity of the piazza in Italian cities.

A piazza often is characterized by its location within the city. Each is unique, and each is given identity by its intended function or by the buildings that surround it. The actual or intended function of the piazza can be categorized as "civic," when located near civic or city buildings, "religious," when associated with a church, or "commercial," when surrounded by shopping centers, cinemas, or restaurants.

Its level of activity determines the success of a piazza. The most successful piazzas tend to be those that are attractively located and offer diverse uses, often defined by the nature of or activity associated with the surrounding buildings. Museums, offices, restaurants, theaters, and hotels are examples of diverse public uses that sustain activity and assure the vitality of the space.

Italy has long served designers and city planners interested in the development of town squares. Many of its cities are those associated with having the "perfect" public squares (Zucker 1973). It is indeed true that the Italians have been masters in the development of outdoor public space, and it is therefore appropriate that we examine some of their great works.

Piazza della Signoria

Florence, Italy

Piazza della Signoria in Florence has served as the civic center for Italy's capital for more than six centuries. It was built to accommodate Palazzo Vecchio (della Signoria), which functioned as the seat of government or city hall since its construction in 1314. In 1434, Cosimo dei Medici, a powerful figure in the Republican government, gained control of the palace, which he later restored. Nearly 100 years later, Cosimo I, as the grand duke of Tuscany, took up residence in the palace, making it the official government headquarters. Today it remains a government building, housing all of the offices of the local city government, including the mayor's office (Figure 6–1).

Piazza della Signoria Time Line

1314	Palazzo Vecchio (della Signoria) is constructed and functions as the seat of government for the city of Florence. The piazza is built to accommodate the palace.
1376	Loggia dei Lanzi, designed by Simone Talenti, is built along one side of the piazza to the northeast of Palazzo Vecchio.
1500s	The two predominant spaces that make up the piazza are organized and defined by the addition of sculptural pieces.
1534	Cosimo I, as the grand duke of Tuscany, takes up residence in the palace and makes it the official government headquarters.
1560–1574	The Uffizi palace, designed by Giorgio Vasari, is constructed to house the administrative body of the Medici state.
1690	A fire destroys areas of the palace.
1792	A major restoration of the palace occurs.
1872	Palazzo Vecchio becomes the property of the city of Florence.
1908–present	The Art Department of the City Council of Florence plans a major restoration of the palace. Restoration efforts continue today.

FIGURE 6–1 The Palazzo Vecchio for which the piazza was built.

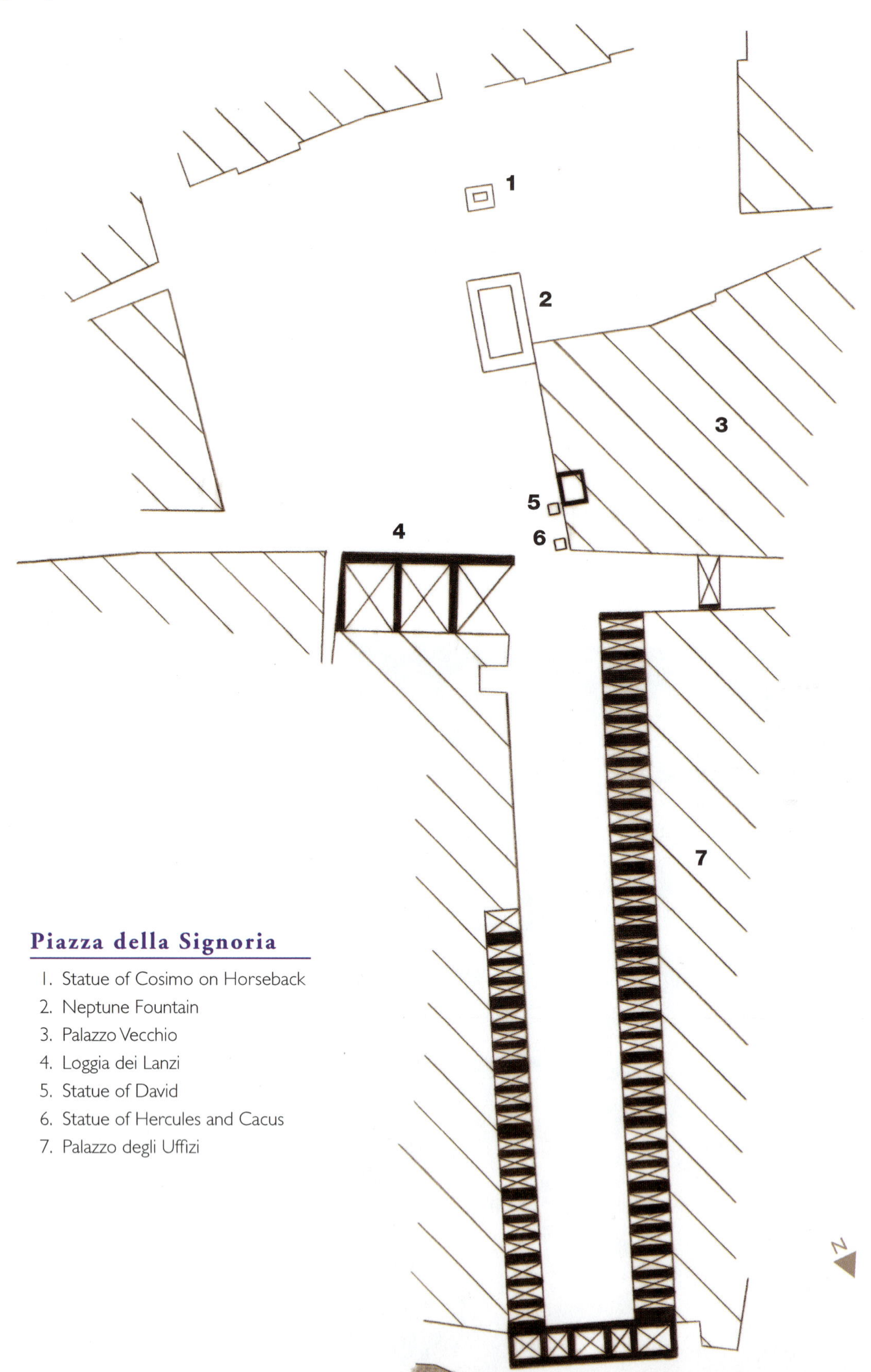
1
2
3
5
4
6
7
N
Piazza della Signoria
1. Statue of Cosimo on Horseback
2. Neptune Fountain
3. Palazzo Vecchio
4. Loggia dei Lanzi
5. Statue of David
6. Statue of Hercules and Cacus
7. Palazzo degli Uffizi

FIGURE 6–2 The piazzetta adjacent to the Uffizi Gallery functions as a passageway from the Ponte Vecchio to the main piazza.

In 1288, Arnolfo de Cambio was commissioned to build the palace for the Priori of the Greater Guilds, a class of merchants and bankers that controlled the city of Florence in the late 1200s. It is a massive building with a fortresslike appearance, presiding over the square. The palace and the square were planned together, each complementing the other and equally contributing to the medieval character of the space. The scale of the building, its architectural details, and even the plaza paving have an austere quality that conveys the spirit of that era.

The piazza was designed as a wide-open area intended for grand political ceremonies and public functions. It is an L-shaped space that surrounds the building on two sides. The main square is thus two distinct spaces. A third space was created as a result of the building of the Uffizi Gallery (1560–1574). This space functions as a passageway from the Ponte Vecchio to the main piazza (Figure 6–2).

FIGURE 6–3 The Loggia dei Lanzi, today an exhibition space, once functioned as a stage for ceremonial events.

The two predominant spaces in the piazza were organized and further defined by the gradual addition of sculpture during the sixteenth century. Renaissance designers, steadfast in their pursuit of spatial order, balance, and harmony, organized the space by creating a visual line between the two areas through the addition of significant sculptural pieces. Michelangelo's David was positioned left of the palace entrance in 1504. Thirty years later, Baccio Bandinelli's Hercules and Cacus was added to the corner of the square, south of the doorway. Located on a diagonal from the corner of the palace is the Neptune fountain by Bartolomeo Ammannati, which was added to the collection in 1575. At the center of this fountain is a large marble statue of Neptune, given the name *Biancone* (Italian for "white one") by the locals because of its light color. In 1594, the last and perhaps most impressive statue was added to the square, the Equestrian Statue of Cosimo I dei Medici. It was commissioned by Ferdinand I dei Medici to honor his father's memory. Its location in the piazza completes the visual boundary that divides the piazza into two distinct areas.

In 1376, the Loggia dei Lanzi was built along one side of the piazza to the northeast of Palazzo Vecchio. This loggia designed by Simone Talenti, was created as a stage for ceremonial functions. Today it is a unique exhibition space. Several fine sculptural pieces are situated beneath the arcade, including Giambologna's Rape of the Sabine Woman (1583) and Ben Venuto Cellini's Perseus (1554) (Figure 6–3).

The last of the three important buildings in this historic civic center is Palazzo Uffizi, by Giorgio Vasari, which was completed in 1580. The Uffizi was built to house the administrative body of the Medici state but later acquired a new function under Grand Duke Ferdinand I. Ferdinand I redesigned the top floor of the building as a gallery to house the family's extensive art collection. Today the Uffizi Gallery has one of the largest public collections of Italian Renaissance paintings in Europe.

Piazza della Signoria embodies the spirit of the Middle Ages while at the same time exhibiting principles of Renaissance theory. In the process of its design, an open, rather formless space, with a medieval appearance, was given shape and proportion without altering its original character.

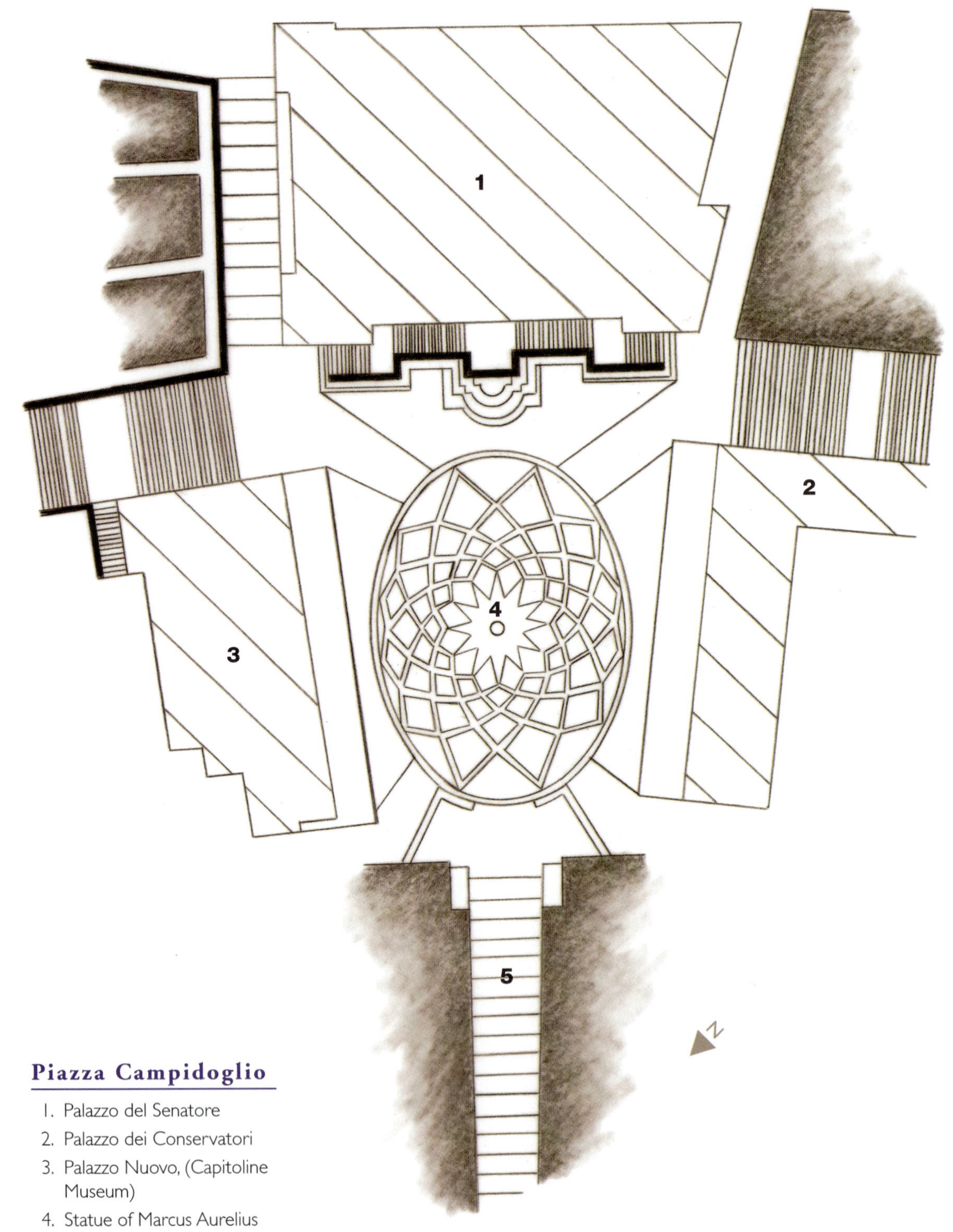

Piazza Campidoglio

1. Palazzo del Senatore
2. Palazzo dei Conservatori
3. Palazzo Nuovo, (Capitoline Museum)
4. Statue of Marcus Aurelius
5. Cordonata

Piazza Campidoglio

ROME, ITALY

Piazza Campidoglio (Campidoglio is the Italian name for "Capitoline Hill") is located in one of the seven hill settlements of ancient Rome. Since the earliest days of Rome, it was an important civic and religious site. Most of its religious association was lost, however, by the Middle Ages, when the settlement gained political importance as the seat of the municipal prefect.

During the early sixteenth century, the site existed as a random assembly of buildings and antique sculptures. Two buildings, originally constructed for the early Roman Senate and Conservatory, occupied the grounds, along with antique statues that had been collected around the city since the late fifteenth century and randomly placed on the site. It was in 1538, during the papacy of Paul III, that work began to create a monumental civic square on Capitoline Hill. The famous artist, sculptor, and architect Michelangelo was commissioned to design a plan to coordinate the randomly assembled space. He was to consider the improvement of the existing architecture and to create ceremonial access to what would become a monumental square commensurate to Campidoglio's significant status.

Michelangelo had two existing buildings to incorporate into the design of the piazza. He retained both buildings but made architectural improvements to them. The first, the twelfth-century Palazzo del Senatore, is the palace of the capitol built on the ruins of the first-century A.D. record office, or Tabularium, and at an angle to the north is Palazzo dei Conservatori, built under Pope Nicholas V (1397–1455) in 1450 atop an ancient temple. Opposite Palazzo dei Conservatori, placed at the same angle to Palazzo del Senatore, Michelangelo designed the newest building, *Palazzo Nuovo* (Italian for "new palace"), to balance the Conservatori, providing symmetry and enclosure to the space. It was completed in 1654 by Girolomo Rainaldi and renamed the Capitoline Museum in 1734. Each building is equally proportioned in plan, but the designed perspective gives Palazzo del Senatore a monumental appearance. Next to the Capitoline Museum is the neighboring Church of Santa Maria in Aracoeli, established in 1250.

The piazza is a small space compared to other public squares built during the Renaissance period, spanning only 181 feet across its widest point. But the space does not appear small, partly because of the sense of movement created by Michelangelo's use of perspective. The plaza is approached from the Cordonata, a monumental ramped staircase envisioned by Michelangelo but constructed by Giacomo della Porta in 1578. The focal point and organizing element of Michelangelo's plan for the plaza is the second-century A.D. bronze equestrian statue of Marcus Aurelius, emperor of Rome (A.D. 161),

Piazza Campidoglio Time Line

1538	Work begins to create a monumental square on Capitoline Hill. Michelangelo is commissioned to create a plan for the piazza.
1564	Michelangelo dies.
1578	The Cordonata, a monumental ramped staircase, is constructed by Giacomo della Porta.
1654	Michelangelo's Palazzo Nuovo, built to create a balanced composition together with the existing Palazzo Conservatori and Palazzo Senatore, is completed. The building is renamed the Capitoline Museum in 1734.
1664	The piazza is completed by Giacomo della Porta.
1940s	The piazza paving envisioned by Michelangelo is finally constructed.
1979	A nearby explosion of a terrorist's bomb leads to the removal of the original equestrian sculpture of Marcus Aurelius from the piazza in 1981.
1990	A replica of the antique sculpture of Marcus Aurelius is returned to Capitoline Hill to replace the original sculpture.

FIGURE 6–4 The bronze statue of Marcus Aurelius in the center of the plaza (the current statue is a replica) was documented as the only surviving statue of ancient Rome.

which stands at the center of the space in line with the center of Palazzo Senatore (Figure 6–4). The sculpture of Aurelius, remembered as one of Rome's greatest rulers, is one of the last surviving monuments of ancient Rome. The statue, now a replica, stands on a base also designed by Michelangelo. Radiating from the statue is a star-shaped paving pattern bound within a large oval form (Figure 6–5). The design, which frames the statue and unifies the buildings surrounding the space, was conceived by

FIGURE 6–5 Radiating from the statue of Marcus Aurelius is a star-shaped paving pattern envisioned by Michelangelo but installed centuries later.

Michelangelo but not actually completed until the twentieth century, dating from the 1940s.

Michelangelo made little progress on the project before his death in 1564. One hundred years passed before the completion of the square by Giacomo della Porta, Michelangelo's successor. For urban designers, Campidoglio is a fundamental lesson in how the constraints of existing conditions—in this case, buildings, statues, and an irregular site—can be integrated into a design without compromising specific program requirements or design objectives.

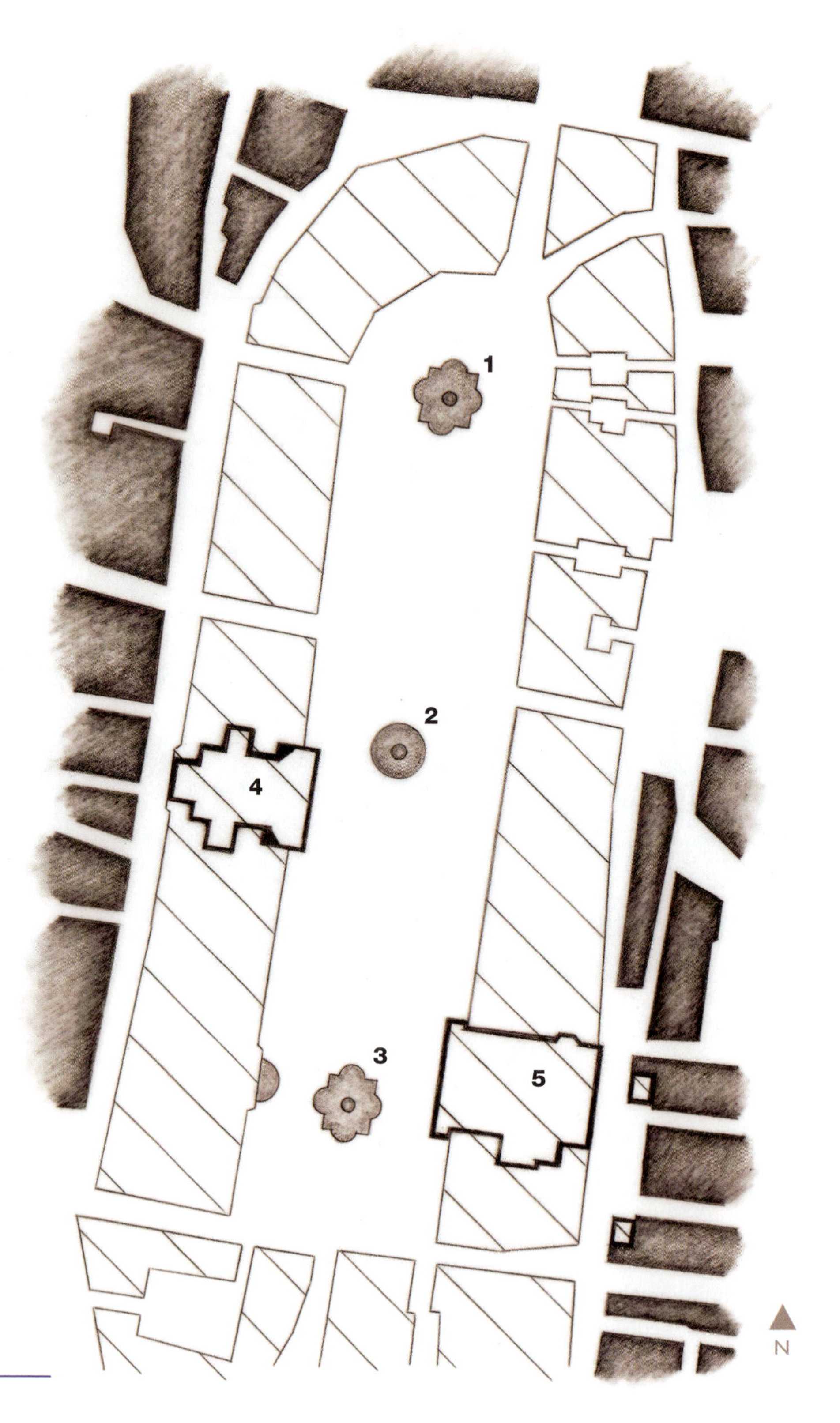

Piazza Navona

1. Moro Fountain
2. Four Rivers Fountain
3. Neptune Fountain
4. St. Agnese Church
5. Church of San Giacomo degli Spagnvoli

Piazza Navona

Rome, Italy

Piazza Navona, located in Old Rome, one of Rome's most beautiful districts, is a long, narrow, U-shaped space converted in the seventeenth century from the Roman stadium built by Emperor Domitian (A.D. 81–96). It is a religious plaza associated with two churches. The oldest, the Church of San Giacomo degli Spagnuoli, dates from the fifteenth century. Opposite this church is the seventeenth-century baroque church, St. Agnese (1652–1677), designed primarily by Francesco Borromini in collaboration with Girolamo and Carlo Rainaldi (Figure 6–6).

One of the most progressive baroque talents, seventeenth-century artist and architect Gian Lorenzo Bernini, worked on the piazza between 1647 and 1651, organizing the space through the addition and revision of fountains. Bernini's most celebrated work, the Four Rivers Fountain (Fontana del Quattro Fiume), was executed between 1648 and 1651. It stands on the longitudinal axis of the square in front of the Church of St. Agnese. It was commissioned by Innocent X, who ascended the papal throne in 1644. Four giant marble figures are grouped around a rocky mound from which an Egyptian obelisk, the symbol of Divine light and Eternity, ascends the sky. Water flows freely from these figures, which personify the four great rivers of the world, the Danube, Nile, Ganges, and Plate. The Danube (of Europe) looks amazed as it gazes toward the obelisk. Both the Danube and the Nile support the pope's emblem composed of lilies and a dove, a symbol of peace and the Holy Spirit. The Nile (of Africa) is depicted with his head shrouded, as its source was yet unknown. To the right of the Danube, the Ganges (of Asia) holds an oar, alluding to the navigable river. Facing the Church of St. Agnese is the Rio de la Plata (of America), looking up with his hand shielding his face (Figure 6–7). The Italians joke that this figure appears horrified by the design of Borromini's church, as the baroque church was indeed a forward statement in architectural design. A more true interpretation is that the figure is protecting his eyes from the obelisk, which symbolizes the bright light of the Christian faith. During the eighteenth century the piazza was often flooded as part of a festive summer celebration. The Quattro Fiume Fountain supplied the water that transformed the piazza into a man-made lake on which boats were sailed as part of a festive summer celebration.

FIGURE 6–6 The Church of St. Agnese.

Piazza Navona Time Line

1633	Bernini remodels the Moro Fountain, originally designed by Giacomo della Porta.
1647–1651	Gian Lorenzo Bernini transforms the Roman stadium built by Emporer Domitian.
1648–1651	Bernini's most celebrated work, the Quattro Fiume, or Four Rivers Fountain, commissioned by Innocent X, is built on the longitudinal axis of the square in front of the Church of St. Agnese.
1800s	The Fountain of Neptune, originally designed by Bernini, is constructed on the north end of the piazza, completing the central spine.
20th c.–present	Restoration begins/continues on several buildings surrounding the piazza.

FIGURE 6–7 A sculpted figure in Bernini's Four Rivers Fountain shields his face with his hand.

Bernini also was responsible for the other two fountains along the center line of the square, the Moro Fountain to the south (Figure 6–8) and the Fountain of Neptune at the north end (Figure 6–9). The Moro Fountain is the southernmost fountain whose original design is credited to Giacomo della Porta. Bernini remodeled the fountain, adding several sculptures to della Porta's work.

The last of the fountains constructed, the Fountain of Neptune, also was designed by Bernini but not executed until the nineteenth century. It stands on the north end of the plaza and completes the central spine.

FIGURE 6–8 The Moro Fountain.

FIGURE 6–9 The Fountain of Neptune.

Bernini's fountains give Piazza Navona its identity. They define the shape of the piazza by emphasizing its length and encouraging movement from one end to the other. Each distinct "fountain zone" becomes a space—a different experience to be enjoyed, almost like a new staged scene in a play. People become a part of the space as they experience it. There is a theaterlike quality to Piazza Navona, a quality sought after and popularized by baroque artists throughout Italy during the seventeenth century.

Piazza San Pietro

1. St. Peter's Church
2. Piazza Retta
3. Piazza Obliqua
4. The Great Colonnade
5. Piazza Rusticucci

Piazza San Pietro
Rome, Italy

For Catholics around the world, St. Peter's Church in Rome symbolically represents the very source of Christ's realm on earth. The Church, the Vatican, and the Piazza are impressively connected as an entire complex of extraordinary scale that celebrates this spiritual center of Christendom.

The various parts of this religious precinct employed some of the greatest talents in both the Renaissance and baroque periods. In 1506, Pope Julius II named Donato Bramante (1444–1514) chief architect for the design of the new St. Peter's Church. Bramante's proposal called for a centralized church based upon the Greek cross, rising above the tomb of St. Peter. Bramante's idea of a central plan, though unique, was rejected in favor of the more traditional and liturgically practical Latin cross design, which naturally provided a length of space for a procession and a hierarchy at the terminus for the location of the main alter.

Over time, the design of St. Peter's Church was addressed by some of the greatest architects of the world. After the death of Bramante in 1514, Giuliano da Sangallo, Fra Giocondo, and Raphael succeeded to his position. Other famous talents who worked on this extraordinary project were Peruzzi, Sansovino, Michelangelo, Fontana, and Moderno.

The great church was completed in 1626, but the associated landscape had been ignored. The various buildings of the complex, with no coherent system of organization, seemed disassociated. Upon the accession of Pope Alexander VII Chigi in 1655, a number of urban projects were initiated throughout Rome. The pope was essentially seeking to create a new master plan for the old city that would improve traffic flow and create opportunities for safe pedestrian circulation. In addition, Pope Alexander focused some of his ambition for city improvements on the Vatican precinct. He commissioned the great baroque artist and architect Gian Lorenzo Bernini to create a public square that would serve as a forecourt to the great St. Peters' Church and provide an approach commensurate to its religious status. The program also required a space that could receive the thousands of visitors who came to worship in the church, attend religious ceremonies and papal appearances, or simply visit this important Christian monument.

In 1656, Bernini developed a plan that organized the vast expanse of space in front of the church by creating three distinct but connected units: Piazza Rusticucci, Piazza Obliqua, and Piazza Retta. The visitor approaching St. Peter's Church from the east (toward the main façade of the church) is first oriented toward the basilica at Piazza Rusticucci. This space lies at the west end of a long avenue, Via delle Conciliazione (Figure 6–10). Although

FIGURE 6–10 The Piazza Rusticucchi collects and directs visitors toward St. Peter's Church.

Piazza San Pietro Time Line

1506	Pope Julius II names Donato Bramante chief architect for the design of the new St. Peter's Church.
1514	Bramante dies, and several other great architects work on St. Peter's Church, including Giuliano da Sangallo, Fra Giocondo, Peruzzi, Sansovino, Michelangelo, Fontana, and Moderno.
1626	The great church is completed.
1655	Pope Alexander VII Chigi commissions a number of urban projects throughout Rome. Architect and artist Gian Lorenzo Bernini is commissioned to create a public square that will serve as a forecourt to the great church of St. Peter's.
1656	Bernini develops a plan for the piazza organized as three distinct but connected units, Piazza Rusticucci, Piazza Obliqua, and Piazza Retta.
1999–2000	The exterior of St. Peter's Church is cleaned and restored in preparation for the Jubilee 2000 celebration of the Catholic Church.

Figure 6–11 Visitors to St. Peter's Church gather in the Piazza Obliqua.

seemingly an unfinished space, Piazza Rusticucci effectively foregrounds Piazza Obliqua, collecting visitors who have approached the precinct from the long east-west avenue that originates at the Tiber. The contrast in moving from Piazza Rusticucci, a much smaller vestibule of space, to the vast expanse of Piazza Obliqua is dramatic and tends to overwhelm most visitors.

Piazza Obliqua, a vast oval enclosure, is the main and central space of Piazza San Pietro (Figure 6–11). Its axis traverses the square with a north-to-south orientation, parallel to the façade of the church. This change in the orientation of axis also causes a change in movement for the visitor approaching the church from the east. An emphasis on the north-south axis, created by the shape of the piazza and the sculptural elements along its center line, slows one's progression toward the basilica. In Bernini's ingenious design for the space, an Egyptian obelisk (c. 1586) by Domenico Fontana for Pope Sixtus V and an existing fountain (c. 1613) by architect Carlo Moderno are used together with a new fountain, similar to Moderno's, in order to reinforce the important crosswise axis. A further design mechanism contributing to the intermediate pause in the visitor's progression toward the church involves the treatment of pavement. The pavement, by its slope to the center of the plaza, together with the radial paving pattern, moves people to the central space, encouraging them to gather as Christians and prepare to enter the sacred monument to their faith.

The piazza's main feature is the twin colonnade composed of columns standing 50 feet high and four rows deep (Figure 6–12). These columns create two majestic arcs that powerfully embrace the assembled crowds. Bernini described them as the "motherly arms of the Church" embracing the faithful in order to reinforce their belief. The colonnade is joined to the front of the church by two corridors that extend from the west ends of the arcs and connect to the outer edges of the church façade. The resulting space, called "Piazza Retta," directs visitors to the main portal. This entire piazza gently

FIGURE 6–12 Surrounding the Piazza Obliqua is Bernini's twin colonnade, which he described as the "motherly arms of the church embracing the faithful."

rises toward St. Peter's vast portico, incorporating a series of monumental steps protruding 80 yards into the plaza. The upward slope of the piazza floor makes the approach to the church entrance extraordinarily powerful.

Piazza San Pietro succeeds as an impressive place of arrival and a masterfully planned assembly ground for the Holy Church. Bernini's ingenious concept is powerful, kinetic, and nurturing all at once. Effective as it is at giving definition to the space surrounding the great church and palace, the emotions that it stirs among its thousands of visitors each year make the piazza even more significant as a part of St. Peter's Church.

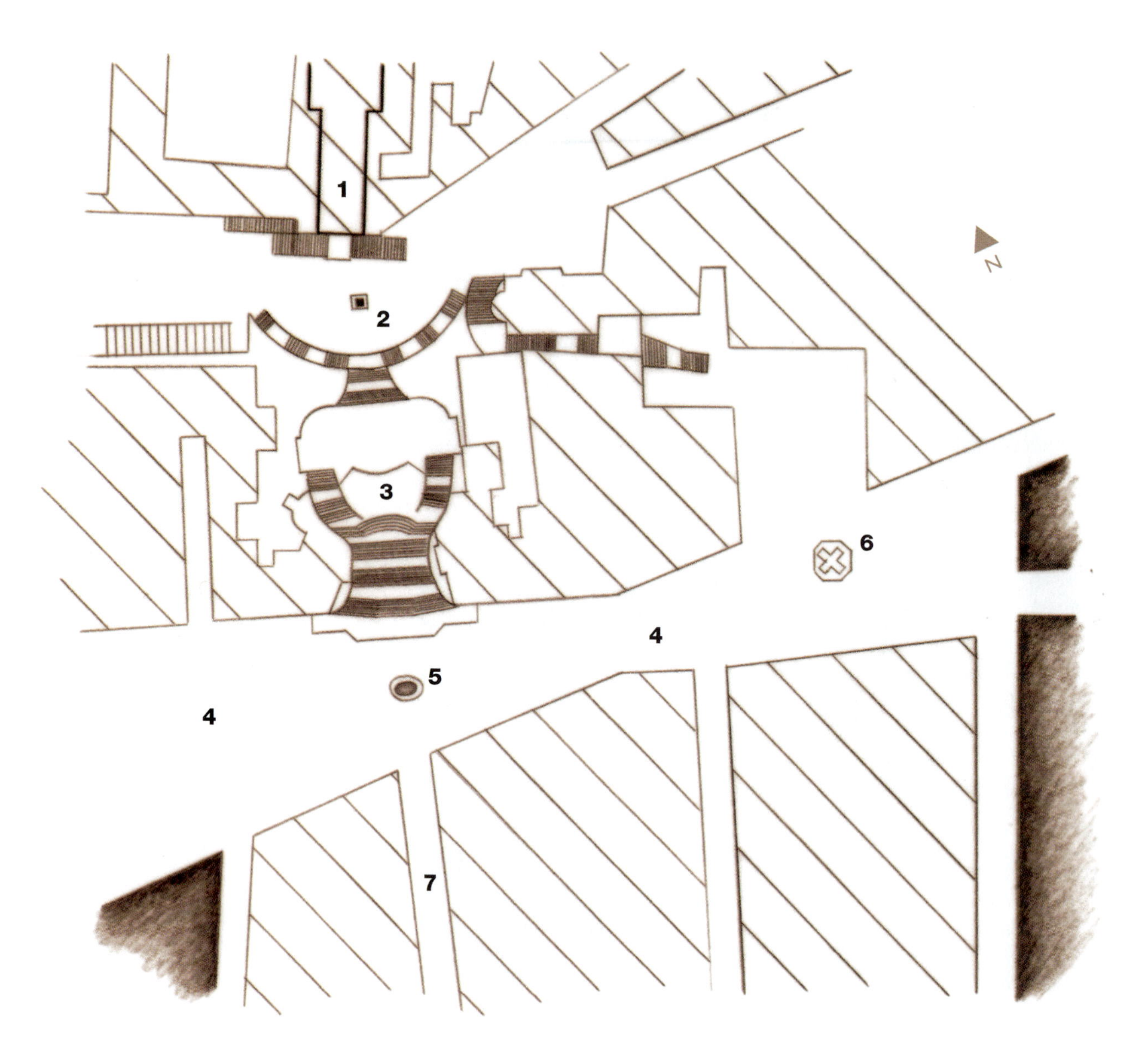

Piazza di Spagna (Spanish Steps)

1. Church of Santa Trinita dei Monti
2. Obelisk
3. Scala di Spagna
4. Piazza di Spagna
5. Barcaccia Fountain
6. Column Commemorating the Immaculate Conception
7. Via dei Condotti

Piazza di Spagna
Rome, Italy

The Scala di Spagna, or "Spanish Steps," are the greatest achievement of Piazza di Spagna. Designed by Francesco de Sanctis and Alessandro Specchi between 1721 and 1725, the piazza, which includes 137 steps, links two topographically different levels in the city of Rome. On the upper level of the site is the Church of Santa Trinità dei Monti (c. 1459). Just below is Piazza di Spagna, which takes its name from the associated palace that was once the Spanish Embassy (Figure 6–13). It has a triangular shape formed by the buildings that surround it. These buildings are interrupted by five streets that converge on the square.

Piazza di Spagna is perhaps one of the most popular social spaces in all of Rome. The square has not gained in popularity but the steps have. The steps have evolved as a dynamic social space, essentially functioning as the square, while the plaza tends to be more ambulatory, filled with people who are either arriving at or departing from the steps (Figure 6–14). The generously wide stairs ascend the hillside from the plaza and are designed to be centered on the church above it. There is actually a slight deviation from the center line of the church, but that is only detected in a plan view

Piazza di Spagna Time Line

1627–1629	Gian Lorenzo Bernini creates a design for the Barcaccia fountain that will be constructed in the piazza at the base of the steps.
1721–1725	Francesco de Sanctis and Alessandro Specchi design the piazza and the 137 steps that link two topographically different levels of the city.
1856	A column topped by a statue of the Virgin Mary is erected at the southern end of the piazza to commemorate the Catholic dogma of the Immaculate Conception. The pope celebrates the Immaculate Conception at that location each year.
20th c.	The streets originating from the piazza evolve into one of Rome's most fashionable shopping districts. Italy's first McDonald's is built nearby.

FIGURE 6–13 A flower vendor brightens a corner of the Piazza di Spagna.

Figure 6–14 People gather on the Scala di Spagna, *or "Spanish steps."*

Figure 6–15 The Barcaccia Fountain, by Gian Lorenzo Bernini.

of the site. The stairway is framed by buildings on either side that seem to contain the space, giving it comfortable scale and a sense of enclosure. The steps are interrupted by several landings that provide a place to rest and also function as viewing platforms from which to observe the bustling activity in the city streets below.

Central to the piazza is a fountain designed by seventeenth-century baroque artist and architect Gian Lorenzo Bernini. The low, broad fountain that lies within an oval basin is sculpted in the shape of a ship and called the "Barcaccia" (Figure 6–15). Like most artful development that occurred during the baroque age, it has a religious connotation. The ship was an honored symbol of the Roman Catholic Church, the "vessel of salvation." Its symbolism derives from the biblical story of St. Matthew, accompanied by Christ and his Apostles in a boat on the Sea of Galilee.

For designers, the Spanish Steps are a fine example of the importance and virtue of recognizing and interpreting the possibilities inherent to a site. Here success was achieved by using existing topography to create one of the most unique outdoor environments in the history of urban and city planning.

PART TWO

France

Environment

France is the third largest country in Europe. It occupies a portion of the European peninsula bound by the Atlantic Ocean to the west, the Mediterranean Sea to the southeast, and the English Channel to the north. Its land boundaries include Spain to the south-southwest and Belgium, Germany, Switzerland, and Italy on its east border, extending north to south. The country is geographically diverse, ranging from very fertile coastal lowlands in the north and northwest, where the greatest population exists, to the extensive mountain region to the east, where historically little settlement occurred because the land is unsuitable for agriculture.

The region's lowlands are dominated by fertile plains that support agriculture production, which accounts for the extent of rural and urban development in this area. The uplands are formed in the south of France and are composed of plateaus, fertile plains, and hills, which become steeper in the southeast region near the mountain ranges. Mountains form the southeastern and southwestern boundaries of the country. To the southeast, the Alps separate France from Switzerland and Italy, and to the southwest, the border between Spain and France is marked by the Pyrenees.

In general, France has a relatively temperate climate, especially in the north and west, where westerly winds carry moisture from the Atlantic over the landscape. This produces a fair amount of precipitation throughout the year and eliminates seasonal temperature extremes. Paris, for example, in the north-central part of the country, receives an average of 188 days of rain each year, with winter temperatures averaging 40°F and summer temperatures 75°F. The south of France experiences milder winters, and the summers are hot and dry. Summer temperatures typically average 10 degrees or so warmer in the south, while winter temperatures are often 10 degrees colder.

The Ile de France has some of the most beautiful and famous gardens in the country, but for these gardens, climate was not the foremost influence in their creation, but rather wealth and power. Most of the gardens in this region have royal connections. Some of the most famous include Fontainebleau, Vaux le Vicomte, Versailles, and Chantilly. Since the sixteenth century, this area of the country was home to a growing community of talented artists. Artists, architects, and gardeners working in this region contributed to France's position as Western Europe's leader in the arts during the seventeenth century. It should be no surprise then that the gardens in this area are some of the best examples of what has been regarded as French garden style. Designed landscapes here were organized on vast areas of level terrain, which became one of the bases for the French garden style.

Central Paris is home to many of the public gardens of France. These gardens or parks have been a part of the city's fabric and urban lifestyle for centuries. Though most are of royal heritage, the royals who owned the gardens opened them to the public centuries ago, and the gardens soon became a very important part of city culture, functioning as a place for community recreation and leisure.

Social and Cultural History

The French nation was incorporated into the Roman Empire in the first century B.C. and was later united with the rest of Western and Central Europe after the Roman Empire collapsed in A.D. 476. During the Middle Ages, several royal dynasties claimed the territory, but their inability to establish an effective governing body over the region gave no legitimacy to their claims.

For many years French society was plagued with religious conflicts and civil strife, lasting through the second half of the sixteenth century. Peace was finally restored by a change of dynasty. In 1598, France came under the rule of Henry IV (r. 1589–1610), the first of the Bourbon kings. Henry's first task was to restore peace to a nation that had been embroiled in political, social, and religious conflict for many years. Under his rule, the monarchy sought absolute power—where the French king alone controlled all government and administration of the kingdom. This concentration of power led to the development of lavish residences and other extravagances conveying the king's supremacy.

Under Henry, Paris was transformed into an exemplary city, and it was established as the royal capital. The king laid plans for an extensive building program that would both improve the city's appearance and glorify him and his nation as the dominant political force in all of Europe. New architectural building projects were initiated, improvements were made to the city infrastructure, which included new roads, bridges, and public spaces, and a grand royal residence was created through the modernization of the Louvre.

Italian precedents inspired much of the work accomplished by French designers during Henry's reign. The king's marriage to Italian-born Marie dei Médicis furthered this Italian influence as the queen sought to create in France what she had so loved in her youth, especially the Italian-style gardens. Grottos, fountains, and sculptures were created in these gardens, modeled after those that adorned many of the splendid villas built throughout Rome and Florence, several of which were owned by the Medici family. From the fifteenth century, the French had been exposed to Renaissance ideas, especially through the Italian Wars (1494–1559), led by Charles VIII. François I (r. 1515–1547) also brought Italian artists to France to contribute to his building projects, the most famous of which was Fontainebleau. He hired some of Italy's finest talents and retained them for several years. The king became one of Europe's leading patrons of the arts beginning in the early decades of the sixteenth century. His building accomplishments were precedents for the great building projects carried out under Henry later in the century.

From 1661, the nature of the monarchy changed with the ascension of Louis XIV (1638–1715), the Sun King. Louis established a new form of absolutism where royal authority was almost godlike. During Louis's reign, there was no limit to the king's power. His aim was to reinforce this image by creating an aura of grandeur about him. He did this mainly through a number of extravagant building projects and gardens, works that not only glorified him but also established France as a leader in the development of the

standards. Several new bridges were constructed, including Austerlitz and the Pont des Arts.

Napoléon made several architectural contributions to the city, most of which were designed to celebrate war victories. The triumphal arch, for example, located at the entrance to the Tuileries Palace, pays tribute to the soldiers who fought in the Austerlitz campaign of 1805, as does the great bronze column erected in the Place Vendôme in 1810.

The fall of the first empire in France marked the beginning of a succession of republics, another empire, and several regimes. The country's instability was a product of distinct political divisions decided during the revolution. Citizens battled each other on issues of the government's structure and economic policies and the Church's role in political activities. Despite all of this internal conflict, France still emerged as a major industrial nation and a leader in the arts and sciences.

Toward the end of the nineteenth century, the continuing progress of industrialization had established a strong and steadily growing economy. This led to a rise in the rate of new construction, with buildings designed by talented, new architects schooled at the Parisian Ecole des Beaux-Arts (School of Fine Arts), established in 1819. This school continued the French tradition of state-sponsored programs for the study of the arts, which had begun in the late seventeenth century with the establishment of the royal academies.

France has maintained its reputation, established during the seventeenth century, as a world leader in the arts. The French style of art, architecture, and landscape, though popularized through royal connections, was not confined to work produced for the French nobility. The lessons of this period's great work rapidly spread to other countries, not simply as an example of style or decoration but rather as a valuable collection of design principles to guide artistic development. The logical and disciplined French approach to design, characterized by precision, elegance, and refinement, gained preeminence as an internationally recognized design style. French classicism continues to inspire many artists and designers who invoke its design methods in their work.

Chapter 7

Elements of French Garden Design

The French garden style has long been characterized as the most formal in its approach to garden design. Since the Middle Ages in France, gardens were created for the wealthy and powerful of society, most often for kings and queens. These gardens were tucked behind castle walls and laid out in neat, geometrically arranged planting beds filled with flowers, herbs, and vegetables. The private and intimate enclosures became more open to the landscape as Italian Renaissance principles began to influence French design. By the mid-sixteenth century, French designers were incorporating their knowledge of the Italian principles of scale, proportion, and spatial organization into their landscapes. At Fontainebleau, François I (r. 1515–1547) employed Italian artists, architects, and sculptors to transform his garden into something resembling the Renaissance style found throughout Italy. Italian principles of design had to be adapted to the French landscape, which was considerably flatter than Italy's. Water, for example, appeared mostly as still pools instead of flowing cascades designed on the hillsides of Italian villas. The climate, too, was quite different, with a greater amount of rainfall and longer periods of cooler weather, making it possible to grow a greater variety of plants than those grown in Italy's hot, dry Mediterranean climate.

During the seventeenth century, the French garden became known for its distinct characteristics, and it emerged as the most popular style throughout Western Europe. Common to the design philosophy of all of the great garden successes in seventeenth-century France was the notion that nature could and should be controlled. The controlling influence of the French monarchy, the greatest patron of the gardens, was clearly evident in what it created. Vast garden plans extended axes to distant horizons, emphasizing power through seemingly infinite perspective. Trees and shrubs were meticulously clipped and arranged in perfect rows and patterns, accentuating the garden's geometry and the extent of human control (Figure 7–1).

FIGURE 7–1 Plants are meticulously clipped, emphasizing the extent of human control (Vaux le Vicomte).

FIGURE 7–2 Parterre de broiderie *gardens at Vaux le Vicomte.*

In 1651, Andre Mollet (d. 1665), a well-respected gardener and writer from a long lineage of French royal gardeners, wrote the book *Le Jardin Plaisir (The Garden of Pleasure).*[1] In it he discussed French theories on and approaches to gardening. He defined the composition of French pleasure gardens by first dividing them into various parts that followed a specific itinerary of progression. First, he noted that "groundworks," referring to low compositional elements, for example, *parterres de broiderie,* were located near the house and never obscured by any tall elements, or "high work," as he called it. This makes perfect sense, as the *parterres de broiderie* were intended to be looked down upon from upper-garden terraces or from the chateau (Figures 7–2, 7–3). Beyond the *parterres de broiderie,* turf *parterres,* called *"parterres a angloise,"* often were created. This entire system, the *broiderie* and turf *parterres,* was typically divided by walkways at right angles to one another.

FIGURE 7–3 A close-up of the parterre de broiderie *at Vaux le Vicomte.*

Water was an essential element in the French garden. Elaborate fountains were designed as focal points, and vast, still pools, known as *mirors d 'eau* (water mirrors), were created to reflect the sky, the house, or even a garden

[1] *Le Jardin Plaisir* referenced in *The Story of Gardening,* P. Hobhouse.

FIGURE 7–4 A tapis vert *along the east-west axis at Versailles.*

FIGURE 7–5 An alleé *of trees frame the road that leads to Vaux le Vicomte.*

building or other interesting feature. Artificial canals were dug to reinforce axes. A main axis was sometimes strengthened through the creation of a wide rectilinear band of turf called a "***tapis vert***" (Figure 7–4). Tree-lined avenues, or ***allées***, reinforced axes and linked garden features (Figure 7–5). Thousands of trees were planted in the garden not only along avenues but also in patterns as garden features, such as the ***quincunx***, laid out as four trees forming a square with a fifth at its center. Other trees were limbed up and pruned to create a sort of aerial hedge that defined garden boundaries (Figure 7–6). Clipped hedges formed architectural structures, and "green" walls, called "**palisades**," enclosed garden rooms. Smaller evergreen shrubs were trimmed in a variety of geometrical forms, adding to a garden's decorative program.

FIGURE 7–6 An allée *of trees on either side of the French pavilion at Versailles forms what is commonly referred to as an "aerial hedge."*

The French, like the Italians before them, united building and landscape. Gardens functioned as an extension of the palace, with individual garden rooms variously designed to create interest. These exterior spaces were every bit as impressive as those within the palace and often quite expansive, with elaborate features and ornate details designed especially for each garden space. There are many examples where gardens were of greater significance than buildings. Chantilly is one such example. In this garden, designed by the great French royal gardener André Le Nôtre, vast areas of water *parterres* and basins are the primary attraction. The main axis is aligned with the garden, not the castle, thus making the castle appear secondary.

Le Nôtre's work, more than any other landscape designer in France, popularized the French style of gardening. His very first project, and certainly one of his finest, was Vaux le Vicomte near the city of Melun, where he experimented with many of the defining elements of seventeenth-century French design. Vaux le Vicomte was followed by Le Nôtre's best-known work and unquestionably the greatest achievement of the French monarchy, the castle and gardens of Versailles. The design approach here was similar to that of Vaux le Vicomte, but on a much larger scale. Its size is unparalleled and represents a significant model to designers for methods of spatial organization on a grand scale.

Chapter 8

Landscape Expression: French Baroque Gardens

As the fifteenth- and sixteenth-century Renaissance style found expression in the landscape and architecture of Tuscany and papal Rome, the seventeenth-century baroque style asserted itself through the creation of great royal palaces and gardens in France. Informed by the Renaissance principles of harmony and spatial order, the French garden developed its own identity as a style of magnificence and perfection. Well-ordered and elegantly defined, the French garden united architecture, art, garden, and site in a unique and masterful way. The French seventeenth-century style, derived from classical aesthetics, flourished as a powerful political statement and display of the dignity and status of France, particularly of the monarchy. As France became the supreme European power of the seventeenth century under Louis XIV (1638–1715), Paris advanced as the capital of art and culture in all of Europe.

Louis XIV assumed control of the government in 1643, asserting his role as a sovereign with absolute power. He formed an intellectual society rich in the arts by institutionalizing those talents in the Royal Academies. These societies were created to support those engaged in the study of fine arts, literature, and the sciences. These three groups were later joined by others to form a larger organization called the "Institut of France," established in 1795. The king's appreciation of the arts was further revealed in the great palaces and gardens he commissioned. These regal estates, monumental in scale and richly appointed, were further proof of France's position as the leading power in Europe. The royal garden played an especially important role in the king's political agenda, providing a venue for entertainment for the aristocratic society, which included many of the king's loyal supporters. These gardens transformed the landscape into an organized ensemble of *allées,* orchards, grand canals, reflection pools, and exquisite fountains. Landscapes reached to distant horizons, along infinite axes, a further expression of the infinite power of the king.

André Le Nôtre (1613–1700) is a name often associated with the French garden style. He established himself as an exceptionally talented designer by transforming the principles of Renaissance garden design to create a new style, a new interpretation of the garden that often has been regarded as the French "formal" style. Le Nôtre's gardening career began early in his life. As a boy he learned the fundamentals of horticulture from his father, Jean Le Nôtre, who was a gardener at Tuileries, the former royal palace built in 1564 for Catherine dei Medici (1519–1589), queen of France. At 16, he entered the studio of Simon Vouet, a painter whose studio became most sought after by aspiring young artists. Despite his early interest in painting, Le Nôtre's passion for gardening would ultimately guide his career. He would succeed his father at Tuilerie and ultimately become the king's master gardener, assigned to create or recreate many of the royal gardens throughout France.

Le Nôtre was a master at organizing space on a grand scale. His work is praised for its splendid composition, clearly united and integrated to form a seamless work of art from many separate parts. His style portrays a clear expression of confidence, unhindered by scale or site conditions. Le Nôtre manipulated the most challenging sites, making them conform to the grand vision he shared with his powerful and royal patrons. Some of Le Nôtre's greatest achievements include the gardens of Vaux le Vicomte, Chantilly, Fontainebleau, and, his most widely known work, the grand royal gardens, developed for King Louis XIV at Versailles.

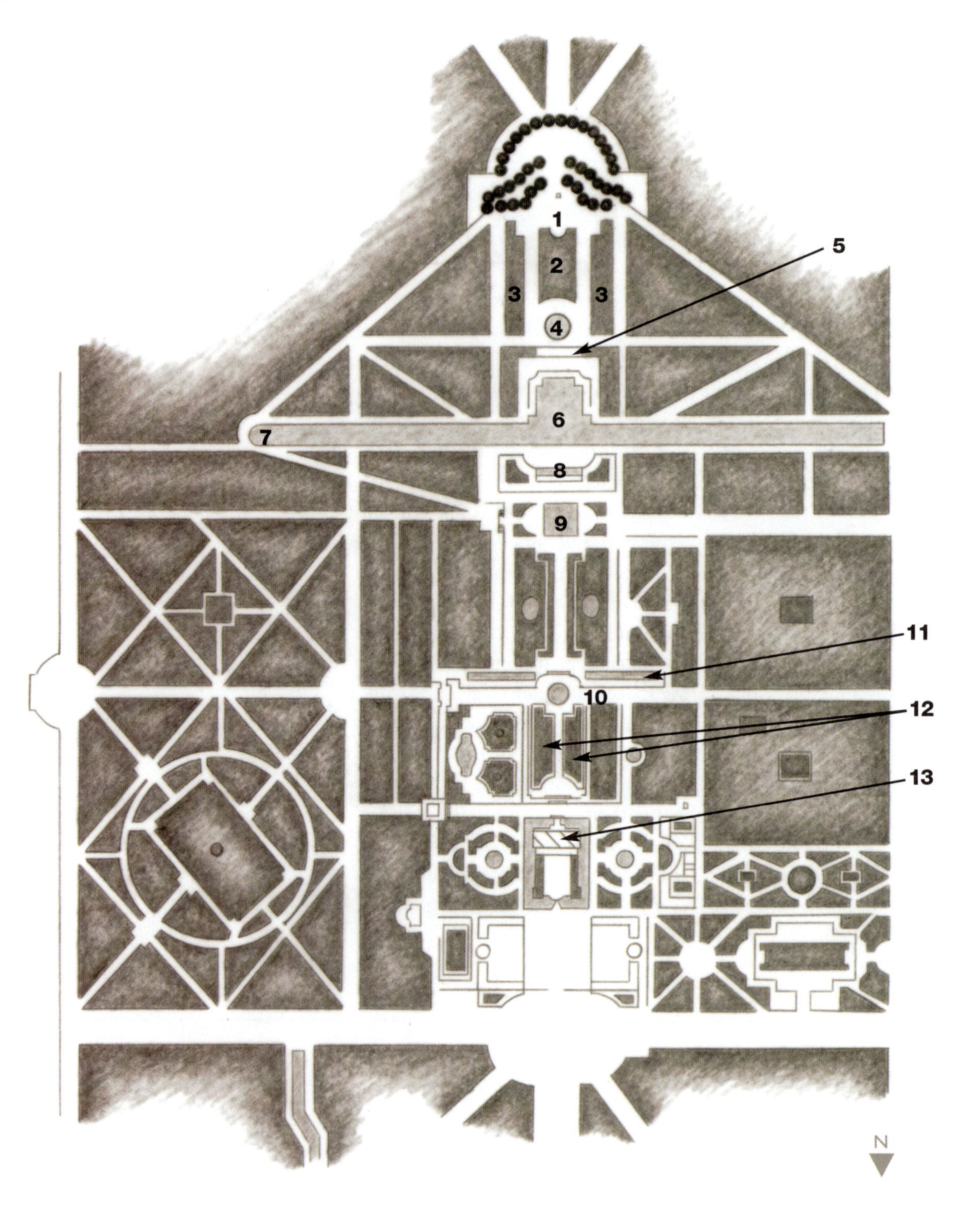

Vaux le Vicomte

1. Statue of Hercules
2. Tapis Vert
3. Bosquets
4. The Gerbe
5. The Grotto
6. The Canal
7. The Poele
8. The Cascade
9. The Mirror of Water
10. The Rondeau
11. The Little Canal
12. The Parterres de Broideries
13. The Chateau

Vaux le Vicomte

SEINE-ET-MARNE, FRANCE

In 1661, France came under the leadership of a new king. Louis XIV (1638–1715) ascended the throne, beginning what would be the longest reign of any king in European history. He ruled the nation for over 72 years, organizing France as the most powerful country in all of Europe. He formed an intellectual society, establishing Paris as the capital of culture, richer in the arts than any other city in Europe. But France's success was not credited to one person alone. Louis XIV surrounded himself with bright, ambitious people. One such person was his finance minister, Nicolas Fouquet (1615–1680). Fouquet was an intelligent, confident man of great wealth and exceptional taste. His famous chateau called "Vaux le Vicomte" was built to represent his fine taste, and it is undoubtedly one of the best examples of French art, architecture, and landscape design of the seventeenth century (Figure 8–1).

Vaux le Vicomte is located 34 miles southeast of Paris. Fouquet acquired the original chateau at Vaux in 1641 and continued to add to the estate by acquiring adjacent properties, anticipating his grand building project.

Nicolas Fouquet's long-term loyalty to the king, the queen mother, and her prime minister, Cardinal Mazarin (who controlled the French government while Louis XIV was a minor), resulted in his appointment to several important government positions early in his life. By 1653, at age 38, he was promoted to the title of "Financial Secretary to the King," which gave him

Vaux le Vicomte Time Line

1641	Nicolas Fouquet acquires the old chateau of Vaux le Vicomte.
1656–1661	André Le Nôtre, Louis Le Vau, and Charles Le Brun design the chateau and gardens.
1661	Fouquet organizes a grand party to celebrate his accomplishments at Vaux and to impress and honor King Louis XIV. The king, suspecting Fouquet of embezzling money to build Vaux, arrests him and imprisons him for life.
1875	Alfred Sommier purchases the ruins of Vaux le Vicomte.
1880	Alfred Sommier begins restoring Vaux le Vicomte. Henri Duchêne and his son, Achille Duchêne, are hired to carry out the work, which continues for over 20 years.
1908	Alfred Sommier dies. His son, Edme Sommier, inherits the estate.
1919	Vaux le Vicomte opens to the public.
1929–present	The chateau and garden of Vaux le Vicomte are recognized as historic monuments. Sommier's direct descendants, Patrice and Christina de Vogüé, continue their work on the preservation of the estate.

FIGURE 8–1 Vaux le Vicomte is regarded as one of the best examples of seventeenth-century art, architecture, and landscape design.

control over the finances of the state. With the news of his promotion, Fouquet set out to build a residence commensurate to his new status. In 1656, he enlisted the services of landscape designer André Le Nôtre (1613–1700), architect Louis Le Vau (1612–1670), and painter Charles Le Brun (1619–1690). Le Nôtre and Le Brun were old friends, having apprenticed together in the studio of Simon Vouet during their youth. The three continued to work on the chateau and gardens for five years.

As the work neared completion, Fouquet organized a grand celebration in the summer of 1661 to recognize his great accomplishment at Vaux. In an effort to impress and honor his king, Fouquet invited Louis XIV and the court to his party, an extravagant event of royal favor. The evening began with guided tours of the interior of the chateau by Le Brun. In the garden, where most of the festivities were to take place, the fountains were gloriously flowing, and the *parterres* were lit with candles in a most elegant display. Guests were entertained with a play written by Jean-Baptiste Poquelin (Molière) and music composed by Jean-Baptiste Lully. Later in the evening, fireworks filled the sky in joyous celebration, but Fouquet would not celebrate much longer, for his lavish display angered rather than impressed Louis XIV. The king, overwhelmed with envy, began to suspect his finance minister of embezzling public funds to build such a great garden. The very next month, Fouquet was arrested, put on trial, and subsequently imprisoned for life. The entire design team at Vaux le Vicomte was immediately sent to Versailles, King Louis XIV's hunting estate, built for his father, Louis XIII, in 1623. There the team would embark on a major transformation that would create a royal palace beyond comparison.

Vaux le Vicomte is a monument to the dignity and status of the garden as an art form in seventeenth-century France. Vaux's garden is organized around long, straight axes, one main central axis and several secondary axes lying perpendicular to it (Figure 8–2). The importance of the garden's central axis is today made evident to the visitor immediately upon arrival in the forecourt, however, early engravings of the garden by Israel Silvestre (1621–1691) and Adam Perelle (1638–1695) reveal that this area is not as it was originally designed. The forecourt was not turfed as it is today. Turf was added later, possibly during the eighteenth century. There was no green space prior to reaching the chateau, or even adjacent to it. The turf *parterres* or **Gazon Coupé** (turf cut into patterns surrounded by colored gravel or sand) panels on either side of the chateau that today form the first zone of the garden are a later addition, likely introduced in the twentieth century.

Three *parterres* that descend in level from west to east dominate the next zone of the garden. The level change is so slight that it is essentially not noticeable until one directly approaches it. The *parterre de broiderie* is central to the space. Here, planting beds of low shrubs and gravel are designed in ornate patterns similar to those created for fine embroidered clothes worn by the society of nobles. To the west is the Parterre de la Courance, or "Parterre of the Crown," named for its central fountain decorated with a gilded crown. This *parterre* is wider than the other two, which is a departure from the symmetrical balance that would have been more typical with such a dominant central axis. The *parterre* occupying the eastern portion of this terrace is the

FIGURE 8–2 The garden is organized around one main central axis and several secondary axes lying perpendicular to it.

Parterre des Fleurs, or "Parterre of Flowers." Seemingly an odd name for a *parterre* composed mainly of turf, this garden was laid out to foreshadow the Trianon gardens that were planted as a brilliant composition of flowers.

The first of the garden's secondary cross axes separates the second and third zones of the garden. Two rectangular-shaped canals lie parallel to this crosswise path to emphasize its east-west orientation. A round pool, the Rondeau, is centered between the two canals in line with the garden's main axis. On the east end of the traverse axis were the vegetable gardens, and on the opposite west end, the *grille d'eau,* a small sequence of waterfalls.

Directly beyond the round pool are steps that descend to the third zone of the garden. This part of the garden is divided in two by a central promenade lined with flower-filled urns, which take the place of what was once a series of elegant water jets that bordered the path on both sides. On either side of the path are rectangular-shaped panels of turf, each having an oval pool positioned at its center. The central path ends at a square pool of water. Just beyond this square pool is the garden's most dominant traverse axis, the Grand Canal (Figure 8–3). It is from this part of the garden that scale is extended, with axes reaching out in all directions. On the north side of the canal, hidden from view by the terrace, is a row of fountains built into a massive wall, together called the "cascades." This elaborate feature is complemented on the other side of the canal by another of the garden's splendid features, Les Grottes. This rusticated stone wall on the south side of the canal is

FIGURE 8–3 The main cross-wise canal.

FIGURE 8–4 The statue of Hercules at the top of the hill, a nineteenth-century addition.

defined by a series of seven niches, each containing sculpted rock work over which water once flowed into a large, square pool. This pool extends out from the crosswise canal, celebrating the arcaded feature and its companion water wall along the opposite bank.

Above the grotto is a round pool, from which an extraordinarily high column of water once rose. It is said that the water reached as high as 16 feet. This feature, now only a simple pool of water, is called the "Gerbe." Beyond this elegant garden feature the main axis was extended into the infinite horizon, along a turf alley, or *tapis vert.* This grassed avenue is flanked on either side by double rows of trees, beyond which are bosquets, meticulously planted on an organized grid. A large figure of Hercules stands on the horizon line at the end of this long green axis (Figure 8–4). It is a nineteenth-century addition that was perhaps considered by Fouquet but did not appear during his time. While this gigantic figure is an exciting attraction, it interrupts the infinite vista to the horizon line common to French design during the seventeenth century. The extension of axes to the distant horizons was a technique that coincided with France's vision of limitless possibilities for both the country and its monarchy.

The gardens of Vaux le Vicomte have long been regarded as one of Le Nôtre's best works, unprecedented in scale and skillfully organized as one grand and united composition. Vaux provides what is perhaps the first example of the French classical style. Le Nôtre assimilated the classicism of the Italian Renaissance, translating its classic principles into a distinctly French language that united architecture, art, garden, and site in a uniquely grand and dignified way.

The chateau passed from the Fouquets to the Field Marshall de Villars, and then to César Gabriel de Choiseul, duc de Praslin, whose family was the last to occupy Vaux before its abandonment, which lasted nearly 20 years. In 1875, Alfred Sommier, a wealthy industrialist, purchased Vaux, which at the time was in a state of ruin. He hired architect Gabriel Hippolyte Destailleurs (1822–1893) to restore the chateau. Landscape designer Elie Laine began the restoration of the gardens under Destailleur. Improvements to the garden were continued by French landscape architect Achille Duchêne, who was hired by Alfred Sommier's son, Edme, when he became master of the estate upon his father's death in 1908. In an effort to raise funds to cover the high costs of maintenance for the estate, Patrice de Vogüé, great-grandson of Alfred Sommier, opened the property to the public in 1967.

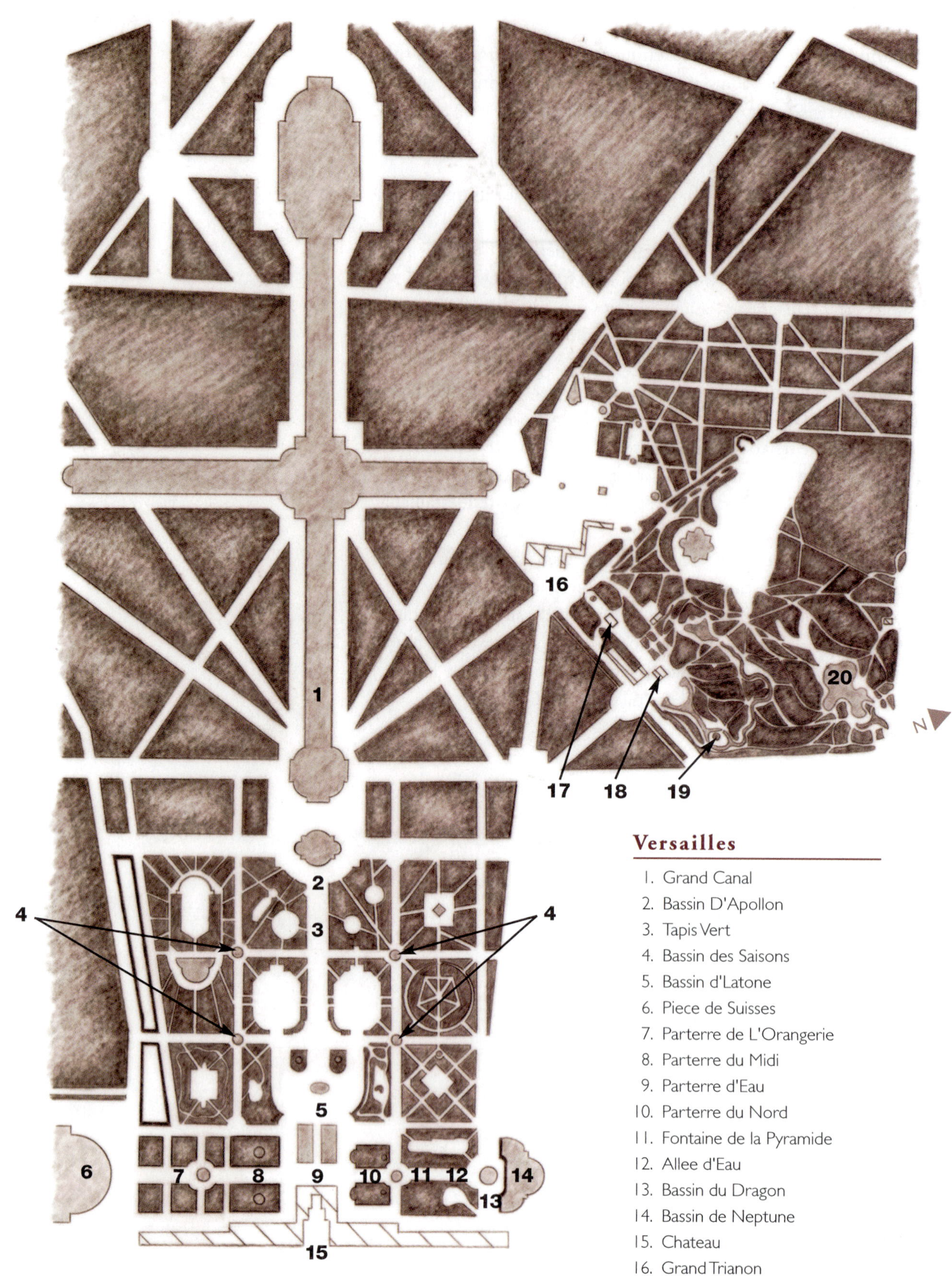

Versailles

1. Grand Canal
2. Bassin D'Apollon
3. Tapis Vert
4. Bassin des Saisons
5. Bassin d'Latone
6. Piece de Suisses
7. Parterre de L'Orangerie
8. Parterre du Midi
9. Parterre d'Eau
10. Parterre du Nord
11. Fontaine de la Pyramide
12. Allee d'Eau
13. Bassin du Dragon
14. Bassin de Neptune
15. Chateau
16. Grand Trianon
17. French Pavilion
18. Petit Trianon
19. Temple de L'Amour
20. Queen's Hamlet

Versailles

Yveslines, France

The story of Versailles is centered on a powerful king and the extent of his life's ambition to create the perfect expression of his royal authority and a symbol of France's power in Europe. The year 1661 marks the beginning of King Louis XIV's personal reign as king. Prior to that date, coinciding with the death of Cardinal Mazarin, Louis had yet to exercise his role as heir to the throne. His mother, Ann of Austria, together with his godfather, Mazarin, had governed the kingdom during his long minority as king, which began at age five. Also in this same year Mazarin's finance minister, Nicolas Fouquet, introduced his newly built chateau, Vaux le Vicomte, to the king—a fatal mistake for Fouquet, but an event that inspired Louis XIV to pursue his extraordinary building project at Versailles. The entire team from Vaux—Louis Le Vau, the architect; Charles Le Brun, the artist and interior designer; and André Le Nôtre, the garden designer—was summoned to Versailles to begin what would be the most extraordinary building project of its time, a remarkable accomplishment even today.

Versailles Time Line

1623	Louis XIII builds a small hunting lodge at Versailles.
1631	Architect Philibert Le Roy designs a modest residence for Louis XIII at Versailles.
1661	Louis XIV becomes king and commissions the building of a grand royal residence at Versailles. He enlists the services of architect Louis Le Vau, artist Charles Le Brun, and landscape designer André Le Nôtre.
1663	Le Vau constructs the first orangery.
1668	Le Vau enlarges the chateau by building an "envelope" around the original house of Louis XIII.
1670	The main elements of the garden are completed by Le Nôtre. The Porcelaine Trianon, designed by Le Vau, is constructed.
1672	The North Parterre, after several enlargements, is completed.
1678	Jules Hardouin-Mansart completes the design for the north and south wings of the chateau.
1678–1687	The Piece des Suisses is created.
1679–1681	The Bassin de Neptune is constructed, balancing the *Piece des Suisses* at the opposite end of the north-south axis.
1682	Louis XIV makes Versailles the official seat of government and the residence of the royal court, totaling nearly 5,000 people.
1684	The new orangery by Mansart is constructed.
1685	Construction of the north wing of the chateau begins.
1687	Construction of the Trianon de Marbre, designed by Mansart, begins.
1715	Louis XIV dies.
1722	Louis XV leaves Vincennes and takes up residence at Versailles.
1762	Architect Ange-Jacques Gabriel designs the Petit Trianon.
1774	Louis XV dies. The reign of Louis XVI begins. Louis XVI orders that the existing trees in the park be cut down and the gardens replanted.
1783	Richard Mique builds the Hameau for Marie Antoinette.
1789	Mobs of revolutionaries arrive at Versailles to seize the king.
1793	Louis XVI is executed.
1805	Napoleon I takes up residence at the Trianon.
1837	King Louis-Phillipe completes the Versailles museum.
1930s	Hameau is restored.
1995	The Establishment Public du Musee du Domaine National de Versailles assumes administrative control of the palace and gardens.
1995–1996	Bosquet de l'Encelade is renovated.
1999	Violent storms destroy thousands of trees in the park.
Present	Interior and exterior restoration efforts continue.

The royal chateau had very humble beginnings. Louis XIII, an avid hunter, built the house as a small hunting lodge in the summer of 1623. Slowly he added to it, creating a more refined residence, yet still rather modest for the house of a royal. It was not unlike many of the country homes owned by the well-to-do people around Paris. Nevertheless, the reserved and unpretentious king always considered it a treasured possession, an affection that would be respected by his much more extravagant son. Louis XIII's completed chateau, designed in 1631 by architect Philibert Le Roy, included a small garden planned by Jacques Boyceau de la Baraudiére and executed by his nephew, Jacques de Menours. Boyceau developed the first garden around a large, central water basin positioned at the crossing point of two wide avenues that divided the garden into four equal spaces. Each of the four resulting spaces was planted with elaborate *parterres.* These highly decorative, embroidery-like compositions were some of the earliest examples of the French-style *parterre de broderie.* The famous French gardener, Claude Mollet, made additional contributions to the garden during a subsequent period of development in 1639.

A quarter of a century later, under the new king, Louis XIV, the gardens would evolve into something extraordinary and monumental in scale. Le Nôtre was responsible for the design. Extending out behind the palace are his famous gardens—the elegant *parterres,* the long avenues, the *tapis vert,* the Grand Canal, and numerous fountains and sculptures. Le Nôtre extended the garden to the farthest reaches of even his own creativeness. He created a system of axes, beginning at the palace with the main axis extending from the center of the chateau west to the distant horizon. The major secondary axis runs parallel to the west facade of the palace, extending north to south. Considerable earth moving, both excavation and filling, was accomplished to create these long, symmetrical axes. Over 30,000 workers moved millions of cubic yards of soil by hand to create the terraced levels along their length. The terraces, aside from being required to create long sight lines, also made the garden much more interesting to experience on the ground. The main elements of the gardens were in place by 1670, when Jules Hardouin-Mansart succeeded his uncle, Louis Le Vau, as the king's architect. That which occurred after 1670, both additions and revisions, resulted in the plan of today.

FIGURE 8–5 A statue of King Louis XIV in the Cour d' Honneur.

The first of the exterior improvements at Versailles was the Great Cour d' Honneur, which opens out before the east front of the palace's façade (Figure 8–5). Adjacent to this court, behind a wide expanse of wrought iron fence and stone balustrades, is the vast Place d' Armes, upon which three long radial avenues, called the "Patte d'Oie," or "Goosefoot," converge. These avenues originate in the adjoining town of Versailles.

On the west side of the palace, immediately beyond the narrow terrace that lies along the central block of the chateau, are two expansive reflection pools, together called the "Parterre d' Eau," or "Water Parterre," which mirrors the building's façade (Figure 8–6). From the western edge of this *parterre* is what can easily be regarded as the garden's most imposing view. Leading

FIGURE 8–6 Laid out before the palace is the Parterre d'Eau, or "water terrace."

from the chateau, along the central and main axis, the sight line continues out over the Basin of Latone, down the *tapis vert* to the Basin of Apollo, and beyond it down the mile-long canal to the distant horizon. At a certain time of year, the solar path is such that the setting sun is directly aligned with this powerful axis, an ingenious design feature and an appropriate counterpart to the opposite end of the sight line centered on the bedroom of Louis XIV, who was known as the "Sun King" (Figure 8–7). Louis compared the power and control of the sun over all objects in the solar system to his own great power as the king of France.

The Fountain of Latone, created in 1670 by Gaspard and Balthasar Marsy, and modified by Mansart between 1687 and 1689, was inspired by the *Metamorphosis* by Ovid. It illustrates the story of Latone, the mother of Diana and Apollo, pleading that Jupiter seek revenge for jeering of the Lycian peasants, which he does by turning them into frogs.

Past the *parterre* of Latone is the *tapis vert,* bordered by bosquets. The bosquets are now considerably overgrown and no longer provide the architectonic effect that a gridded network of trees once offered. However, the groves still succeed as a framing device for the next major sculptural feature, the Basin of Apollo, and also for the Grand Canal, just beyond it.

FIGURE 8–7 View from the king's bedroom window down the mile-long canal.

FIGURE 8–8 The king likened himself to the sun god, Apollo, the central figure in the imagery of Versailles.

At the end of the *tapis vert* is the Basin of Apollo, at the center of which is a gilded lead sculpture of Apollo emerging from the water in his chariot, drawn by four horses and preceded by trumpeting tritons and sea creatures (Figure 8–8). The king likened himself to the sun god Apollo for his bravery, strength, and wisdom. Beyond the Basin of Apollo, the central axis stretches out to the distant horizon along the mile-long canal. The digging of the Grand Canal began in 1667 but was later extended. When Le Vau enlarged the chateau on the garden side, the view from the upper terraces seemed to foreshorten the axial view. This is what led to the extension of the canal to a distance of nearly one mile in order to bring the composition into appropriate scale.

From the Parterre d' Eau, on the west side of the palace along the traverse axis to the south, are three significant features. The first, adjacent to the chateau, is the Parterre du Midi, or "South Parterre." This garden originally was designed as twin panels of splendid embroidery, each having a rather simple but elegant round pool at its center, all against a field of crushed, colored aggregate. The colorful tapestry of flowers filling the garden today was not added until the reign of Louis XV, whose wife, Queen Marie, desired flowers to look upon from her apartment above. It was then that the garden took on the name Parterre a Fleurs or "Floral Parterre." Beyond the Parterre du Midi two grand flights of steps, known as the Cent Marches, or "Hundred Steps," descend to the Garden of the Orangery (Figure 8–9). This garden extends out from the orangery building as a vast area laid out as six broad turf panels, with a large circular pool central to the space. The building, designed by Jules Hardouin-Mansart between 1684 and 1686, replaced a smaller orangery designed by

FIGURE 8–9 Orange, lemon, and palm trees are set in large planters in the Garden of the Orangery.

Louis Le Vau in 1663. During the spring and summer seasons, hundreds of orange trees and palms are set out in the garden, bringing it to life and filling the air with a pleasant fragrance. Immediately to the south of the Garden of the Orangery is a large lake, the Piece des Suisses, which was designed as a large reflection pond to mirror the orangery. Excavation of the lake began in 1678 with an army of men borrowed from the Swiss Guard, thus acquiring the name Piece des Suisses, or "Swiss Lake." The task of dredging the swampy bottomland proved to be more than was anticipated. Many men lost their lives during its construction, being struck down by disease from the mosquito-infested sludge. It was finally completed in 1687, reaching its full size of nearly 30 acres.

To the north of the central terrace, on the same secondary axis, one descends several steps to the Parterre du Nord, or "North Parterre." The earliest version of this garden was developed in 1663, adjacent the north front of the chateau. It was enlarged to nearly twice its original size by 1666, and by 1672, its composition had been significantly altered. Several new water basins and sculptural pieces were designed into a series of triangular turf sections bordered by a collection of carefully clipped topiary. At the north end of the central walk that divides this garden stands the Fontaine de la Pyramide, or "Pyramid Fountain," by Girardon, a four-tiered composition of marble basins supported by tritons and sea creatures cast in lead (Figure 8–10). From the fountain the path descends a steep slope flanked on either side by a series of small water basins supported by bronze sculptural groups representing young children. This section of the path is the Allée des Marmousets. It also is referred to as the Allée de Eau, or "Water Avenue." The axis terminates at the

FIGURE 8–10 The Pyramid Fountain stands along the central walk that divides the north parterre.

bottom of the avenue with a semicircular pool, the Bassin du Dragon, named for the Dragon Fountain at its center. The large dragon rises from the still water surrounded by sea creatures and cupids aboard swans. From its mouth, a stream of water rises nearly 100 feet.

The northern end of the traverse axis terminates at the Bassin de Neptune. This semicircular pool was designed by Le Nôtre and constructed between 1679 and 1681. It succeeds in creating a balance between the opposite ends of the axis. Before the construction of the Neptune Basin, the axis was more heavily weighted to the south end, especially after the establishment of the large lake, the Piece des Suisses. This balance is more apparent in plan

FIGURE 8–11 At the crossing points of the avenues that divide the groves are the Fountains of the Seasons. Spring is represented by Flora, the goddess of flowers.

view, but it is perhaps less obvious from the ground. The shape of the pool, also repeated in the tree-lined boundary that surrounds it, echoes the same form designed for the south pond, again contributing to both balance and harmony between both ends of the sight line. Various sculptural groups, including Neptune, the centrally featured sculpture, are positioned along the south wall of the pool. Also along the top of this wall is a series of large urns with water jets that produce a crystal-like screen of water. This watery partition frames the space and also provides an elegant backdrop to the Dragon Fountain when it is approached from the Allée de Eau.

When most of the work along the two principal axes of the garden was either complete or planned, the designers turned their attention toward the development of the bosquets, or groves, which were divided by a grid of *allées,* one pair oriented north to south, the other east to west. Fourteen different compositions were created among these planted woods. Each of the spaces was uniquely designed, but all were generally created to accommodate the king's busy entertainment schedule. Dancing, dining, music, and plays were among the activities enjoyed by the king and his invited guests in these open-air chambers carved out of the woods. Designers, sculptors, and builders worked together to create glorious stage sets of the finest materials—marble pavilions, gilded lead statuaries (Figure 8–11), and wrought iron trellises—all ready-made for the festivities known as *fêtes*. It was here that the king could display his fine taste and assert his princely status with the pleasures commensurate to wealth and power.

Most of the garden features in the groves have deteriorated beyond repair, but of those remaining, the Garden of Enceladus is perhaps the most impressive. This garden takes on a rather unique octagonal shape, partially bordered by a handsome **arbor** (plants interwoven and trained over a trellis structure) that helps define its boundaries. At the garden's center is its main feature, a colossal figure of the giant Enceladus from Greek mythology, half concealed in a pile of rubble—the result of his failed attempt to scale Mount Olympus, despite the warnings of Jupiter (Figure 8–12). From his mouth a towering stream of water reaches to the sky. He appears to be struggling to save his own life. Perhaps this is yet another symbol alluding to the king's controlling power. A message to anyone daring to challenge the king, it provides a clear example of possible retribution.

At its very beginnings, Versailles was a retreat, a place where the king could escape from the public eye. By 1682, it had grown to be something quite different. It was now the chief residence of a new king and his court of several thousand, a place built to awe and impress the public rather than

FIGURE 8–12 The gilded Fountain of Enceladus.

avoid it. Soon King Louis XIV realized the need to occasionally remove himself from the royal pomp and circumstance. He acquired the village of Trianon, adjacent to Versailles, where he established a private getaway to enjoy with selected friends (and mistresses). The entire village was demolished, making way for the Porcelain Trianon, a small, private residence designed in 1670 by Le Vau. Its name was derived from its exterior décor of blue and white ceramic tiles that appeared much like the blue and white porcelain being imported from China at the time. It was built for the king's mistress at the time, Madame de Montespan, who remained his favorite for more than a decade.

Michele Le Bouteux, a floriculturist and the husband of Le Nôtre's niece, largely directed the Trianon gardens. For this garden the king, who was especially fond of flowers, sought a colorful display of exotic plantings. Le Nôtre, known to care little for flowers of any sort, trusted this work to Le Bouteux. Nearly 100,000 potted plants were rotated in the garden throughout the growing season, yielding a very dynamic design for the pleasure of the king and his invited guests. Orange, lemon, and pomegranate trees were lined out for the summer months, producing fragrant blossoms that were used to decorate the royal apartments.

When Madame de Montespan fell from the king's grace in 1687, his new mistress, Madame de Maintenon, appealed to the king for a different chateau. The Trianon de Porcelaine was replaced by the Trianon de Marbre, a splendid building faced with pink Languedoc marble, designed by Mansart (Figure 8–13). Le Nôtre was responsible for the design of the gardens. In his plan, he retained Le Bouteux's work but also incorporated a more natural-looking scheme. He had earlier tried this technique at Versailles for one of the bosquets, but it was destroyed to make way for Mansart's pink marble peristyle, called the "Colonnade." The destruction of that garden always infuriated Le Nôtre, so his decision to create the same type of composition is not

FIGURE 8–13 The Trianon de Marbre.

surprising, especially for Mansart's new building. It was the perfect opportunity for him to avenge his young rival, who was expecting something much more formal and refined.

When the king died in 1715, his great-grandson and successor, Louis XV, came to live at Versailles. He was responsible for several changes to the garden. Most noticeable was the more natural looking appearance of the garden. Very gradually, the tight-clipped trees and shrubbery were allowed to grow naturally, freed from the constraints put upon them by the king's team of gardeners. We can speculate about the reasons for this change in the care of the garden—little interest, lack of resources, different tastes—but none of these is certain. What is certain is that the garden assumed a very different look. Early-eighteenth-century engravings by Jacques Rigaud reveal the garden's changed appearance.

Another change that Louis XV initiated at Versailles was the construction of the Petit Trianon. In 1762, Madame de Pompadour, the king's mistress, persuaded him to build a small chateau near the larger building constructed several years earlier by Mansart. This building, developed by architect Ange-Jacques Gabriel, was designed in the neoclassical style. Considerably smaller than the first Trianon, this intimate chateau became the queen's favorite place to stay. Its gardens were particularly important to the king, an avid plant collector, who laid out a sizable botanical collection that grew to over 4,000 varieties, one of the largest in Europe at the time.

As the years passed, devastating storms and hard winters caused considerable destruction to the gardens. Many trees were either physically damaged, in poor health, or dead. When Louis XVI ascended the throne in 1774, he made the decision to cut down all of the trees and replant the gardens. Debates followed among his advisors about the style the new gardens should take. The classical tradition of French design was regarded as boring in many circles, having been replaced by a more natural style popularized by the English. The king, with respect for his predecessors, who so thoughtfully created this physical expression of royal authority, made the decision to replant the garden in an identical scheme. Nearly 14,000 trees were planted to replace those that had been diseased or cut down. Le Nôtre's design for the original gardens was strictly adhered to.

While in the main gardens the king did little to express his own tastes, he made significant changes to the Trianons. The king's wife, Marie Antoinette, had grown especially fond of the Petit Trianon, spending considerable time there with close friends. She had a theater built near the chateau to satisfy her desire for theatrical entertainment, often playing a part in the performances. In 1780, the queen participated in *Devin du Village,* a story about the humble lives of peasants and farming families. Envious of this free-spirited and carefree life, she persuaded the king to construct a make-believe farming village for her where she could assume the role of a poor farm maiden set apart from the high society that gathered around the royal residence. The village, called the *Hameau,* was composed of 12 different buildings, constructed in 1783 by Richard Mique, who also was responsible for the construction of the queen's theater. Among these buildings was a dairy,

FIGURE 8–14 One of the 12 buildings constructed in the make-believe farm village created for Marie Antoinette.

a barn, a mill, and several dwellings (Figure 8–14). A real working farm was set up nearby to legitimize the existence of the new buildings. There gardens were laid out and tended, fields were planted and harvested, and animals were kept, all which greatly amused the queen. The landscape around the village was regarded as the first true example of an English-style garden in France. Its agrarian nature and freedom of composition were hailed as a success for this new developing style.

It was in the queen's most favorite garden, on a cool October morning in 1789, that she received the frightening news of the approaching mobs of revolutionaries arriving at Versailles to seize the king. This was the beginning of a major change in the social and political system of France. The absolute monarchy was replaced with a series of different governments, and many new laws and privileges favoring the free and equal rights of France's citizens were established.

Shortly after the king's execution in 1793, the properties were confiscated and made open to the public. The gardens were altered to make commercial use of the grounds, generating income by converting land into orchards, vegetable gardens, and pastures. Forests were timbered, and the wood was distributed to the army and navy.

Napoleon I arrived in 1805 and took up residence at the Trianon. He made only minor improvements to the overall gardens, spending most of his time and money refurbishing the Trianons. The next royal to have a significant impact on the residence was Louis-Phillipe, king of France from 1830 to 1848. He was responsible for transforming the palace into a museum, to glorify France and all of its successes. The gardens were restored, but with little accuracy, replacing gravel paths with lawn and introducing tree varieties inconsistent with the original plantings.

Today, Versailles is a garden filled with visitors from all over the world, totaling nearly 10 million each year. It is supported by the generosity of those drawn to its rich history, a generosity that extends beyond the French. In fact, the majority of foreign visitors to Versailles are American, the people intrigued by its dramatic story.

American foundations have generously funded many of the restoration programs. During the 1990s, the park suffered its worst devastation in centuries. Early in the decade, a storm toppled more than 3,000 trees in the park and caused considerable damage to the gardens. However, the worst destruction occurred in December 1999, when 105-mile-per-hour winds ravaged the park, destroying over 10,000 trees. Through the generosity of many American families and foundations, such as the John D. Rockefeller Jr. family, the Kress Foundation, and the World Monuments Fund, tens of millions of dollars have been raised to fund restoration projects, including the cleanup and replanting from the 1999 devastation.

The long-range plan for the park includes the reestablishment of the main palace gardens to reflect the precise appearance of the scheme created by Le Nôtre during the seventeenth century. It further provides for the restoration of the queen's hamlet, perhaps the area worst-hit by the storms, as well as the Trianons. The projected date of completion is 2015.

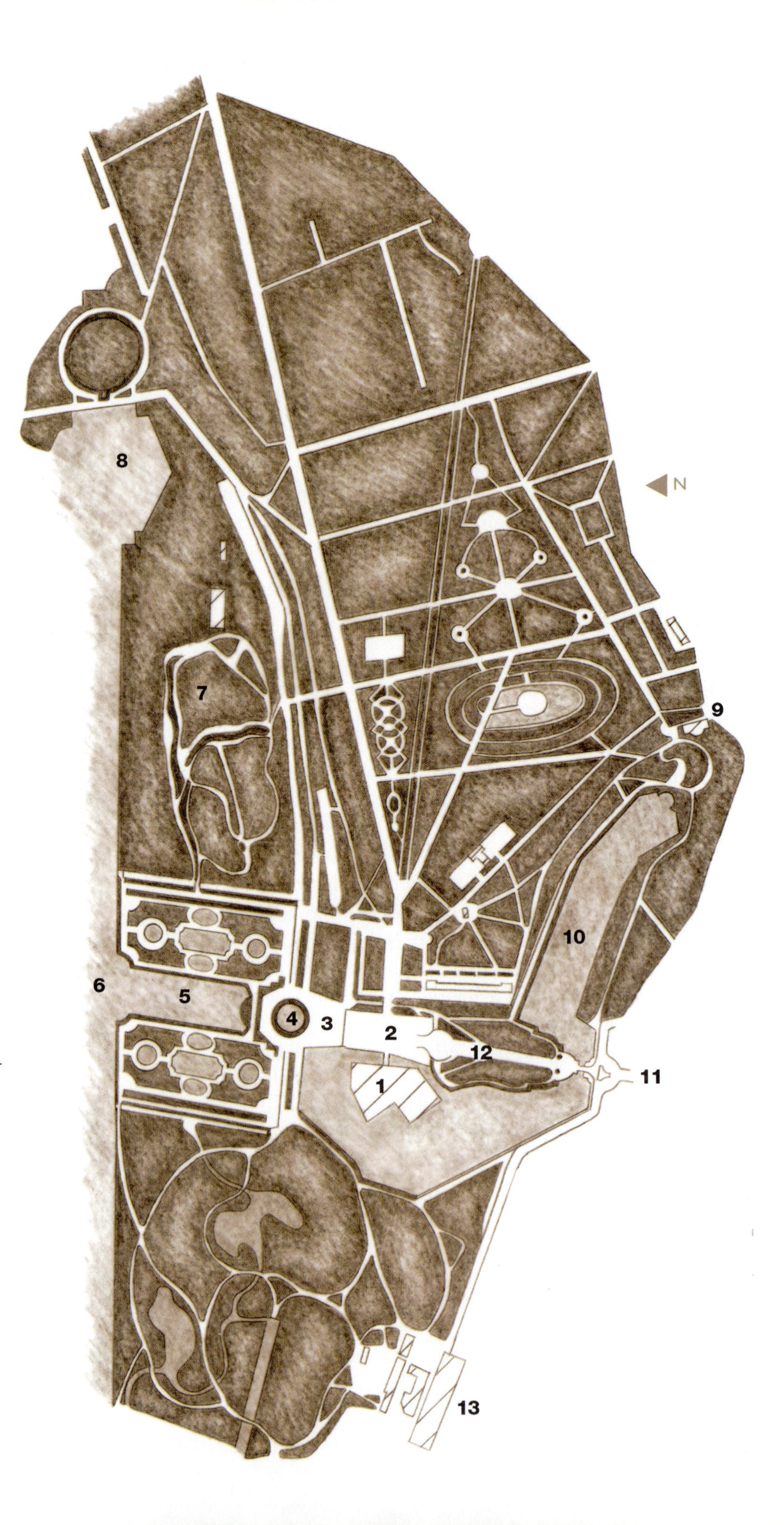

Chantilly

1. Grand Chateau
2. Terrase du Connetable
3. Grand Degré
4. Basin de la Gerbe
5. The Manche
6. Grand Canal
7. Hameau
8. Head of Grand Canal
9. Maison de Sylvie
10. E'Tang de Sylvie
11. Grille d'Honneur (Main Gate)
12. Forecourt
13. Grand Écuries

Chantilly

VAL-D'OISE, FRANCE

Chantilly is one of the finest creations of the French Renaissance. It has been called "the most beautiful house in France,"[1] shared by a great number of men and women who passionately built, rebuilt, and embellished it for several centuries. The finest architects, artists, and landscape gardeners of the day were commissioned to work there. The duc d'Aumale, the last of a long line of royals associated with the estate, called it "a monument of French art in all of its branches."[2]

Chantilly Time Line

10th c. Charters identify the Chantilly estate as first belonging to the Bouteillers of Senlis.

1386 Chantilly is sold to Pierre d'Orgemont, chancellor to Charles V, who commissions the reconstruction of the chateau.

1484 Pierre d' Orgemont leaves Chantilly to his nephew, Guillaume de Montmorency.

1531 Guillaume de Montmorency leaves Chantilly to his son, Anne de Montmorency. Anne makes improvements to the estate, increasing its size by adding several buildings.

1567 Anne de Montmorency dies.

1614 Henri II de Bourbon, prince de Condé, inherits the estate through his marriage to Charlotte Margauerite, the last of the family's heirs.

1632 Henri II de Montmorency is condemned to death after being found guilty of participating in a rebellion against the king's chief minister, Cardinal Richelieu. King Louis XIII confiscates Chantilly.

1643 King Louis XIII dies. The young duc d'Enghien, soon to be prince de Condé, defeats the Spanish at Rocroi. Ann of Austria, queen regent, returns Chantilly to the Condé family.

1654 Condé is condemned as a rebel for leading a campaign against Chief Minister Cardinal Jules Mazarin. Chantilly is again confiscated.

1659 Condé is pardoned by Louis XIV, and Chantilly is returned to Condé.

1660 Condé sets out to make improvements to the estate.

1662 French landscape architect André Le Nôtre is hired by Condé to transform the garden at Chantilly.

1686 The Great Condé dies. His son, Henri-Jules de Bourbon, continues to embellish the chateau and grounds.

1710 Louis-Henri, duc de Bourbon, the seventh prince de Condé, great-grandson of Henri-Jules de Bourbon, inherits Chantilly and sets out to make improvements.

1721 Architect Jean Aubert creates the Écuries for Louis-Henri, duc de Bourbon.

1740 Louis-Henri, duc de Bourbon, dies. Louis-Joseph de Bourbon succeeds him at Chantilly.

1762 The prince de Condé, a distinguished French general, leads the royal armies during the Seven Years' War.

1789 The revolution breaks out, and Louis-Joseph de Bourbon and his son, the second duc de Bourbon, emigrate to safety in Flanders.

1792 The chateau of Chantilly is sacked.

1811 Under the Imperial Regime, led by Napoleon I, Chantilly is given to Napoleon's step-daughter Queen Hortense.

1814 Napoleon I abdicates. Chantilly is returned to the Condé family upon their return to France when the first restoration begins.

1815 The prince de Condé returns to Chantilly and sets out to restore the estate.

1818 The prince de Condé dies. His son, Louis-Henri Joseph, the duc de Bourbon, has little interest in the restoration of Chantilly.

1830 The duc de Bourbon dies and leaves the estate to his nephew and godson, Henri-Eugene, duc d' Aumale.

1844 The duc d' Aumale takes up residence at Chantilly.

1848 Revolution breaks out in Paris. The duc d'Aumale flees to England in exile with his father, King Louis-Phillipe.

1871 The republican regime abolishes the laws of exile, and the duc d' Aumale returns to France and to Chantilly. The duc d'Aumale is elected to the Académie Française.

1876 The duc d' Aumale sets out to restore the palace.

1884 The duc d' Aumale creates a will and names the Institut de France heir to Chantilly.

1897 The duc d' Aumale dies.

1898 The Musée Condé opens to the public.

1988 The Foundation Électricaté de France restores the statues in the park.

[1]Statement made by Henri IV, noted in Chantilly, Babelon, 1999.
[2]Chantilly, Babelon, 1999.

The palace is located 25 miles from Paris on the outer limits of the Ile de France. Here many leading figures who were influential in the decisions of the monarchy once resided. Charters from the late tenth century refer to a place known as "Chantilly," which at that time in history belonged to the Bouteillers of Senlis. From its earliest beginnings, the chateau was no more than a fortress. It would not change until it came into the possession of Pierre d'Orgemont, chancellor to Charles V, in 1386. The d'Orgemont family, who held the estate well into the fifteenth century, made great improvements. The last of the d'Orgemonts died without an heir and left the property to a nephew, Guillaume de Montmorency, who immediately took up residence at the estate. He became one of the most respected aides to monarchs Louis XI, Louis XII, and Francois I, and it was through these royal associations that Chantilly assumed a new level of notoriety. It began to assume a more regal aspect, coinciding with its growing popularity among leading figures and monarchs.

Guillaume de Montmorency (d. 1531) left the property to his son, Anne de Montmorency (1493–1567), constable of France (1552) and a respected advisor to several kings. Anne made a number of significant improvements to the estate and increased its overall size by adding several buildings and reconstructing others. Chantilly was passed down through the Montmorency family to the last of the family's heirs, Charlotte Margauerite de Montmorency (1594–1650), who was married to Henri II de Bourbon, the prince de Condé (1588–1646), and also was the mother of Louis II de Bourbon, known as the "Great Condé" (1621–1686).

The Great Condé, a cousin to Louis XIV (1638–1715), king of France, admired Chantilly like no other before him. With him, the chateau gained the reputation of a most magnificent home with the finest appointments, rivaled only perhaps by the palace of the king (Figure 8–15).

Figure 8–15
The palace.

Chantilly was especially dear to Condé, as Louis XIII (1601–1643), during his reign, had taken it away from the family for several years when Henri II de Montmorency was found guilty in 1632 of participating in a rebellion against Cardinal Richelieu. As a result, all of the properties owned by the family, including Chantilly, were confiscated. It would not be until the death of the king that a decision would be made for their return. Ann of Austria (1601–1666), the queen regent, was responsible for that decision, and in fact her motivation to return the properties was owed to Condé. Condé, the young duc d'Enghien, 21 years old, had just saved France by defeating the Spanish at Rocroi. The regent, seeking to honor him for his heroism, decided to return Chantilly to the family. The estate would be lost again when Condé himself was condemned as a rebel after leading a campaign, called the "Fronde," against the monarchy beginning in 1648. More than a decade would pass before the property was returned to Condé as part of a pardon granted to him by the king. Condé expressed his loyalty to Louis XIV and became one of his principal courtiers.

Two years after taking up residence there, Condé set out to make improvements to the estate. He began with the garden. In 1663, he hired renowned landscape gardener André Le Nôtre, who had just completed work on the garden of Vaux le Vicomte, which was hailed as the greatest achievement in landscape design in all of France at that time. Le Nôtre began transforming the gardens at Chantilly, while at the same time developing plans for the landscape surrounding the great royal palace of Louis XIV at Versailles. At Chantilly he was assisted by his nephew, Claude Descots, and architect Daniel Gitard, who also had worked at Vaux prior to Le Nôtre's arrival. Their work evolved over a 20-year period. Gradually Chantilly would become a splendid composition of long, extended vistas, skillfully proportioned terraces, and vast, still pools that constitute the garden's uniqueness and its overall spirit.

Additional land was acquired adjacent to the property to allow Le Nôtre to carve out the long vistas that he envisioned. A *bois vert,* or "green wood," which was quite fashionable at the time, was created to the west of the chateau to enhance the prince's view from his apartments. To the north, the valley of the Nonette was filled in order to create a major axis in that direction. A circular basin, the Basin de la Gerbe, was constructed along the axis and later complemented on either side of the centerline by large *parterres* developed just north of the pool. From 1672 to 1673, Le Nôtre drained the meadows and diverted the river. Its flow was channeled into a huge canal, constructed across the entire site. The canal extended into the north-south axis to provide a place where boats could be launched. To the west, Le Nôtre designed a series of formal terraces, later replaced by an informal garden in the English fashion. On the east side, a garden pavilion was restored and complemented by a grassy park carved out from the adjacent forest. But perhaps the most grand of Le Nôtre's creations at Chantilly was the main approach that he created to the chateau and gardens from the adjacent forests. Confronted with a rather awkward system of converging lanes, Le Nôtre decided to redesign the approach, creating a long, straight avenue, the Allée

FIGURE 8–16 The Allée de Connetable.

FIGURE 8–17 The equestrian statue of the Constable is the focal point of the approach axis.

Connetable, which led into the estate from the forest (Figure 8–16). The focal point of this new approach axis would not be the chateau but rather a statue of Condé's ancestor, Anne de Montmorency, the former constable and a leading statesman in the royal kingdom (Figure 8–17). The initial perspective culminates with the equestrian statue of the constable, which stands high on the grand terrace at the point at which the house and garden are united. From the statue, a full view of the garden is exposed, which was previously hidden from view in the initial perspective by the rising grade of the approach avenue. Once one reaches the Grand Terrace, the statue, once a focal point, is transformed into the main observation point into the gardens. On the south side, a long, wide ramp from the forecourt to the Grand Terrace adds to the monumental character of the approach, while on the north side, a series of majestic steps descends to the garden. These steps, called the "Grand Degré," were designed by Gitard. Le Nôtre and sculptor Jean Hardy are credited with the creation of the grottoes beneath the staircase.

Condé retired to Chantilly and died there in 1686. His son, Henri-Jules de Bourbon, continued to embellish the chateau and the grounds, but it was his great-grandson, Louis-Henri, duc de Bourbon (1692–1740), minister of Louis XV, and the seventh prince de Condé, who contributed substantially to Chantilly over a 30-year period, from 1710 to 1740. It was he, with a passion for hunting and a habit for spending, who constructed the famous Écuries, or stables, an enormous building constructed to keep as well as to display his fine collection of horses and hounds (Figure 8–18). After the death of Louis-Henri, duc de Bourbon, in 1740 there was very little new construction at Chantilly. His son and heir to the estate, Louis-Joseph de Bourbon (1736–1818), the last prince de Condé, was then only a four-year-old child. In 1741 his mother died, making him an orphan. His inheritance was managed by his guardian uncle until 1748, when the young prince took personal possession of the estate.

Condé brought pride to the family, rising to the ranks of a distinguished French general and leading the royal armies during the Seven Years' War. But when the revolution broke out in 1789, he and his son, the second duc de Bourbon (1756–1830), emigrated to safety in Flanders. The princes had escaped the trials of war, but Chantilly did not. By the summer of 1792, the chateau had been sacked, the furnishings and collections taken, and several buildings, including the Grand Chateau itself, demolished. Under the Imperial Regime, led by Napoleon in 1811, Chantilly was given to Queen Hortense, Napoleon's stepdaughter, but was soon returned to the Condé family upon their return to France when the first restoration began in 1814.

The prince de Condé set out to restore the estate, recovering the property that had been sold to others and thus reasserting its original boundaries.

FIGURE 8–18 The Écuries, or stables, built by Louis-Henri duc de Bourbon to keep and display his collection of horses.

The recovered areas to the west, where Le Nôtre had developed a series of terraced *parterres,* were replaced with informal gardens laid out in the newly developing parklike fashion made popular by the English. This area accordingly became known as the Jardin Anglais, or "English garden."

FIGURE 8–19 A sculpture of dramatist Moliére, added to the garden during the nineteenth century.

The Prince de Condé died four years after his return to Chantilly. His son, the duc de Bourbon, really had no interest in continuing the restoration of the chateau begun by his father, preferring instead to spend most of his time there hunting in the forest. When he died in 1830, he left the estate to his nephew and godson, Henri-Eugene, duc d' Aumalé (1822–1897). Aumale moved to Chantilly with his new bride in 1844. As soon as they settled in, they began to make plans for a major reconstruction of the chateau. These plans, however, were suddenly halted when a revolution broke out in Paris in February 1848, leading to the deposition of King Louis-Phillipe (1773–1850), the duc d' Aumalés' father. The entire family fled to England, giving up all immediate plans for Chantilly's rebirth.

The republican regime abolished the laws of exile in 1871, and the duc d' Aumale returned to France and to military service, where he reassumed his rank of general. By 1876, having lost both his wife and his children prematurely, he turned his attention once again to the reconstruction of the palace. His building program included several large reception rooms, a grand formal dining hall, residential suites, and a private chapel. That same year he also was elected to the Académie Française. The duc d' Aumale was a man of scholarship with an appreciation for the arts. During his exile in England, he had amassed a sizeable collection of books and paintings. As part of the new design for the chateau, gallery rooms and a library were created to store and display those collections.

FIGURE 8–20 A sculpture of André Le Nôtre, added to the garden during the nineteenth century.

In the garden, the duc d' Aumale restored the Grand Parterres created by Le Nôtre and continued to develop the English gardens to the west of the site. He commissioned artists to create sculpted figures for the garden of those who had contributed to Chantilly's long legacy, especially during the period when the Great Condé occupied the chateau. Statues were created of people such as dramatist Moliére (1622–1673) (Figure 8–19), who added to the spirit of the place through the many performances he gave for Condé and his invited guests, and garden designer Le Nôtre (Figure 8–20), who contributed greatly to the atmosphere of Chantilly through his impressive design of the elaborate water gardens surrounding the chateau.

France, by 1883, was guided by the republic, and the duc d' Aumale began to question the fate of his palace and his vast collection of art upon his death. In an effort to preserve his life's collections for all of France to enjoy, he left his entire estate to the Institut de France. The collection is now housed in the Musée Condé, a name chosen by the prince to commemorate the noble family to which he was born and by which he inherited Chantilly. The museum contains the original collection of nearly 1,000 paintings, including works by Raphael and Salvatore Rosa, several thousands of drawings and engravings, and sculpture, furniture, and tapestries. The library is home to over 30,000 literary works. It is indeed one of the most unique and interesting collections in the French artistic heritage.

Fontainebleau

SEINE-ET-MARNE, FRANCE

For eight centuries, Fontainebleau has been a royal home and hunting ground to many of France's greatest kings. From its earliest history in modern times it contributed to the spread of Renaissance art in French culture. Talented artists and painters who were trained in Italy were employed at Fontainebleau. These artists appropriated Renaissance principles of art, architecture, and gardening, reinventing techniques to create a style with its own identity, one suited to aristocratic culture, particularly the monarchy. Fontainebleau was the scene of royal hunts and lavish celebrations, of marriages and births, and of artistic and cultural appreciation and learning.

The palace of Fontainebleau is located in a town of the same name. It is situated in the north-central region of France, just southeast of Paris near the valley of the Seine River. The chateau's history can be traced back to 1137, when it was mentioned in a royal charter of King Louis VII, but it really begins in the first half of the sixteenth century with François I (1494–1547). Upon his return from captivity in Spain, he sought to reestablish his court in Paris after more than a century away in the Loire valley. It was the beautiful forests that drew François to Fontainebleau, one of the largest woodland areas

Fontainebleau Time Line

1137	Earliest mention of the chateau of Fontainebleau in a royal charter of King Louis VII.
1528	King François I reestablishes his court in Paris and settles at Fontainebleau, where he begins restoration on the house and gardens.
1535	The Jardin des Pins is created.
1547	François I dies, and he is succeeded by Henri II. Henri II hires French architect Philibert Delorme as superintendent of works.
1559	Delorme designs the Great Horseshoe Staircase. Henri II dies. Francesco Primaticcio replaces Delorme as the superintendent of royal buildings, under Catherine dei Medici.
1589	Henri IV becomes king and settles at Fontainebleau.
1594	Henri IV has the Jardin de l'Étang built.
1599	Henri IV commissions the redesign of the Grand Jardin and the Diane Garden.
1610	Louis XIII becomes king.
1632	The Great Horseshoe Staircase is redesigned and built by architect Jean Androuet du Cerceau.
1643	Ann of Austria, queen regent, hires André Le Nôtre to redesign the Jardin de Diane and the Grand Formal Garden.
1715	Louis XV, as king, commissions the demolition of the interior of the Belle Cheminee, or Beautiful Fireplace Wing, in order to construct a court theater.
1789	The French Revolution brings an end to the royal occupation of the residence until the reign of Napoleon I in 1804.
1804	Napoleon I, as emperor of France, takes up residence at Fontainebleau.
1830	Louis-Philippe becomes king and sets out to completely restore Fontainebleau.
1852	Naploeon III becomes emperor of France and continues to carry out the restoration work on the palace begun by Louis-Philippe.
1940	The German army occupies Fontainebleau during World War II.
1979–1986	The Musée Napoléon is created.
1986–present	Restoration of pavilions and apartments continues.

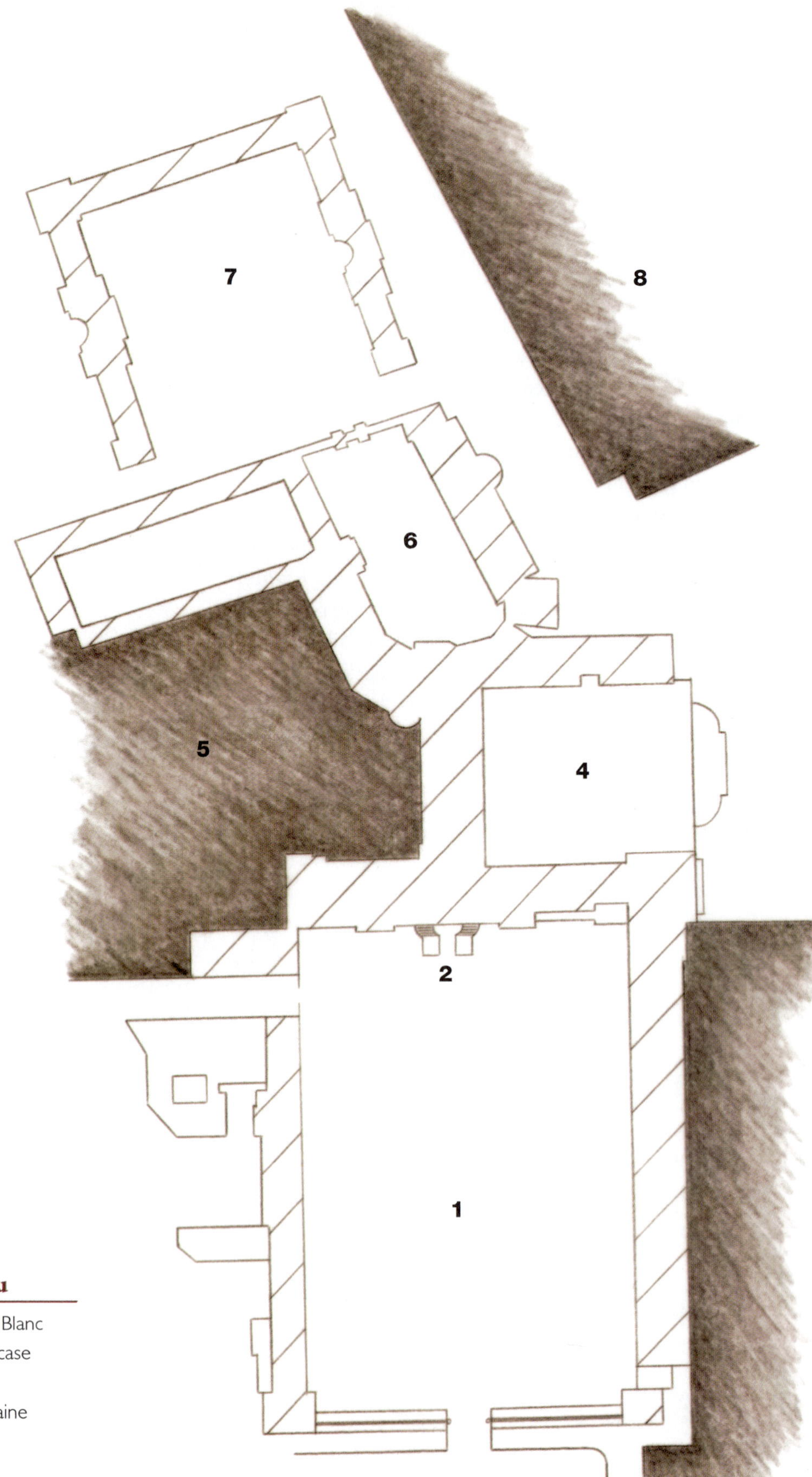

Fontainebleau

1. Cour du Cheval Blanc
2. Horseshoe Staircase
3. Jardin Anglais
4. Cour de la Fontaine
5. Jardin de Diana
6. Cour Ovale
7. Cour des Offices
8. Grand Jardin

in the Ile de France, at that time covering over 35,000 acres. The forest of Fontainebleau had been known for its natural beauty and large population of wild game, especially deer. Its many acres of wooded terrain provided an exceptionally great hunting ground for the king and his invited guests. In addition to wild game, the forest was an economic resource, providing acres of available timber.

In 1528, François began a series of architectural improvements around the city, including the château de Madrid in the Bois de Boulogne, the Château in Saint Germain, and Fontainebleau, where the greatest amount of work occurred. When François arrived at Fontainebleau, it was in a state of ruin. He immediately set out to restore the castle to make it habitable.

The king had a keen taste and an appreciation for the many Italian masterpieces created by the great Renaissance artists and architects working in Italy at the time. He was fortunate to acquire some of those same talents for Fontainebleau. The sacking of Rome in 1527 was unfortunate for Italy, destroying much of the Renaissance progression, but it proved a great opportunity for the king and for France. Many Italian artisans left Rome during this period and found work in France, giving birth to ideals of Renaissance art and architecture that would soon develop a new expression in a different culture and landscape.

In 1530, Florentine artist Giovanni Battista di Jacopo, also known as Rosso Fiorentino, arrived at Fontainebleau, followed by Francesco Primaticcio from Bologna two years later. The transformation of the chateau into a splendid achievement of Renaissance reconstruction was due in part to these talented men. Other Italian artists contributing to the project included painter Luca Penni, who arrived from Florence in 1530, sculptor Giralomo della Robbia, also from Florence, who arrived in 1518, and Florentine sculptor Domenico del Barbiero and architect Giacomo Barozzi da Vignola from Modena, both employed at Fontainebleau under Primaticcio after 1540. Perhaps the most famous Italian architect to work at Fontainebleau was Sebastiano Serlio, who arrived at the royal castle in 1541. He would spend the rest of his life there working as a painter and an architect to King François.

The work of many of these artists defined what has been regarded as the developing style of **mannerism** (developed from the Italian Renaissance of the early sixteenth century using classical forms but in ways that were more powerful, emotional, and elegant). Rosso Fiorentino is credited with founding the so-called School of Fontainebleau, which was essentially a manifestation of the mannerist style, derived from the work of Italian artists contributing to Fontainebleau from 1530. The work of Italian-trained French artist Toussaint Dubreuil and artists Ambroise Dubois and Martin Fréminet, both of France, represented what is called the "Second School of Fontainebleau," which coincides with the reign of Henri IV.

Throughout his reign, King François continued to make improvements on the buildings, beginning with the royal apartments in the Cour Ovale (the Oval Court), then the François I Gallery wing, a covered walk linking the royal apartments and the monastery. Alterations on the chateau

FIGURE 8–21 The Cour Ovale, or "Oval Court."

continued throughout the 1530s. By the end of the decade, the king had developed a model royal residence and had given rise to the Renaissance movement in France.

After having made several improvements to the buildings, his attention turned to the surrounding landscape. In 1535, he established a garden, the Jardin des Pins (the pine garden), where he planted maritime pine trees in an unsuccessful attempt to recreate an Italian garden. The garden was designed around a rustic grotto, the first of its kind in a French landscape. The grotto survived, but unfortunately the pines did not, and the garden ultimately was redesigned in later years. It seems that the king enjoyed this garden, as what followed was the popular Cour du Cheval Blanc (White Horse Court) (Figure 8–21), which created an architectural link between the garden and the chateau.

Of the kings who followed François I, few had as great of an impact on the garden as he, with the exception of both Henri II (1519–1559) and Henri IV (1553–1610). Henri II, the son of François I, also was quite fond of Fontainebleau and continued work on the chateau after the death of his father. When Henri came to the throne in 1547, he hired architect Philibert Delorme, a Frenchman, as superintendent of works, making him responsible for all new construction relating to the royal residences. In this period of development we see a shift in the origin of talent at the royal residence, with a greater number of French artists employed by Delorme, albeit lesser-known talent. This was perhaps due to his personality, which has been described as rather difficult and authoritative. His first task as superintendent of the royal

FIGURE 8–22 The Horseshoe Staircase, situated before the entrance to the royal apartments.

buildings was to make repairs and improvements. In addition, he built a cabinet, or an apartment, for the king, and he created a ballroom between the oval courtyard and the king's garden. But perhaps one of the most impressive of his works was the Great Horseshoe Staircase, a monumental set of stairs before the entrance to the royal apartments (Figure 8–22). Delorme created this design for Henri just before the king's death in 1559.

Francesco Primaticcio (1504–1570), an Italian painter and architect, replaced Delorme as superintendent of royal buildings under Catherine dei Medici (1519–1589). He continued at Fontainebleau until his own death in 1570. One of his most significant accomplishments at the palace during his tenure as superintendent was the construction of the Aile de la Belle Cheminee (the Beautiful Fireplace Wing). This building, with its monumental entrance, gave the royal home a new level of prestige. It also increased the significance of its associated courtyard. The courtyard, known as the Cour de la Fontaine (Fountain Court), is named for a fountain on its east end that featured a sculpted figure of Hercules created by Michelangelo early in his career, circa 1504. This figure would later be replaced by a sculpture of Perseus under the direction of Henri IV in 1594. Primaticcio also was responsible for creating a

FIGURE 8–23 The Fountain of Diana, in the garden that bears its name.

moat around the royal residence in 1565. This was done under the direction of Queen Catherine, who sought to make the residence a safer place, since it was no longer simply a pleasure home but an official royal residence that housed the royal court.

Henri IV (1553–1610) was the next king to contribute substantially to the royal residence. After restoring peace within France at the end of the Wars of Religion (1560s), Henri settled at Fontainebleau, where he made several contributions despite his short period of reign. He made improvements in the *Cour Ovale* with the addition of a monumental gateway, the Porte du Baptistere (the Baptismal Font Gate), so named for the baptismal event that he held there in 1606, the christening of the future Louis XIII. Henry also commissioned the building of several additional outbuildings to accommodate an increasingly larger royal court.

King Henri IV preferred not to appoint any single person in charge as chief architect. Instead, he spread the work between several artists and architects. The most talented among them constituted what would be called the "Second School of Fontainebleau," which echoed the first, but with a predominance of French talent.

In making significant additions and alterations to the garden, Henri enlisted the services of highly talented French landscape designer Claude Mollet (1563–1650). Mollet collaborated with an Italian hydraulics engineer, Thomas Francine. Together they redesigned several of the gardens at Fontainebleau, including the Grand Jardin and the Diane Garden. Francine designed elaborate fountains for each—the Tiber Fountain in the Grand Jardin and the Fountain of Diana for the garden that bears its name (Figure 8–23). Henri also was responsible for the Jardin de l'Étang, an artificial island built in 1594, which is linked to the main garden by a footbridge. Standing atop this island is a stone pavilion that essentially dominates the space (Figure 8–24).

The flurry of building activity at Fontainebleau ended with the death of Henri IV. His son, Louis XIII, had little interest in the arts and architecture,

FIGURE 8–24 A stone pavilion dominates the Jardin de l'Etang, an artificial island built in 1594.

FIGURE 8–25 The Grand Formal Garden, designed by Le Nôtre and Le Vau.

thus there was limited new construction during his reign. Projects begun by his father were seen to completion, while existing site features in a state of decline were reconstructed. The Great Horseshoe Staircase, for example, originally designed by Delorme, was redesigned and rebuilt between 1632 and 1634 by architect Jean Androuet du Cerceau (1632–1634), but only because it was close to collapsing. Upon Louis's death, his wife, Ann of Austria, queen regent from 1643 to 1661, made several improvements to the interior of the palace as well as to the gardens. She had the Jardin de Diane completely redesigned by the soon-to-be famous André Le Nôtre (1613–1700). Le Nôtre also was responsible for the design of the Grand Formal Garden, which he worked on with architect Louis Le Vau (1612–1670) under the direction of Louis XIV, who assumed personal reign in 1661. Le Nôtre and Le Vau created a vast *parterre* garden with a large water basin at its center that contained 34 fountains (Figure 8–25). Aside from these interventions, Louis XIV did little to change Fontainebleau, as most of his attention was focused on building the great royal palace at Versailles.

The next sovereign to have an impact on Fontainebleau was Louis XV, although much of his work has been the subject of negative criticism. A strong shift in tastes and desires in the early eighteenth century led to some drastic changes, such as the complete destruction and redesign of the interior of the beautiful fireplace wing to create a court theater. Besides these interior changes to the residence, entire buildings also were torn down, such as the Galerie d' Ulysee, built by François I, which was replaced by a larger structure that was not in balance with adjacent buildings. Many of these changes were regretted by subsequent generations.

The French Revolution, lasting from 1789 to 1799, brought an end to the royal occupation of the residence for several years. The chateau was damaged during the war, but fortunately the damage was not irreparable. It was not until the occupation of Napoleon I, who became emperor of France in

1804, that Fontainebleau was restored, making it fit once again for a sovereign to occupy. Napoleon was fond of the palace and did his best to restore it to its former condition. He also made changes to the garden, particularly the Jardin des Pins, redesigning it in the popular English fashion. His architect, Maximilian Joseph Hurtault, accomplished this between 1810 and 1812. The garden was subsequently renamed the Jardin Anglais (English garden).

During the nineteenth century, King Louis-Philippe (1773–1850), a man who was dedicated to the arts and had a keen interest in the history of the palace, set out to completely restore it. The existing Renaissance ornamentation was refurbished, and new floors and ceilings were installed, in keeping with the style of the period.

Napoleon III (1808–1873) continued to carry out the restoration begun by Louis-Philippe until the fall of the Second Empire. It was not until the mid-twentieth century that another major restoration project was carried out. Since then, several work projects have been accomplished through various programs that support restoration operations.

Fontainebleau is an exciting medley of various styles and periods coexisting in a unique and noteworthy composition. It is the story of tradition, of a shared fondness by kings for a small chateau in the midst of a forest that evolved into a most admired residence of the French royal court.

Chapter 9 Landscape Expression: Modern Variations

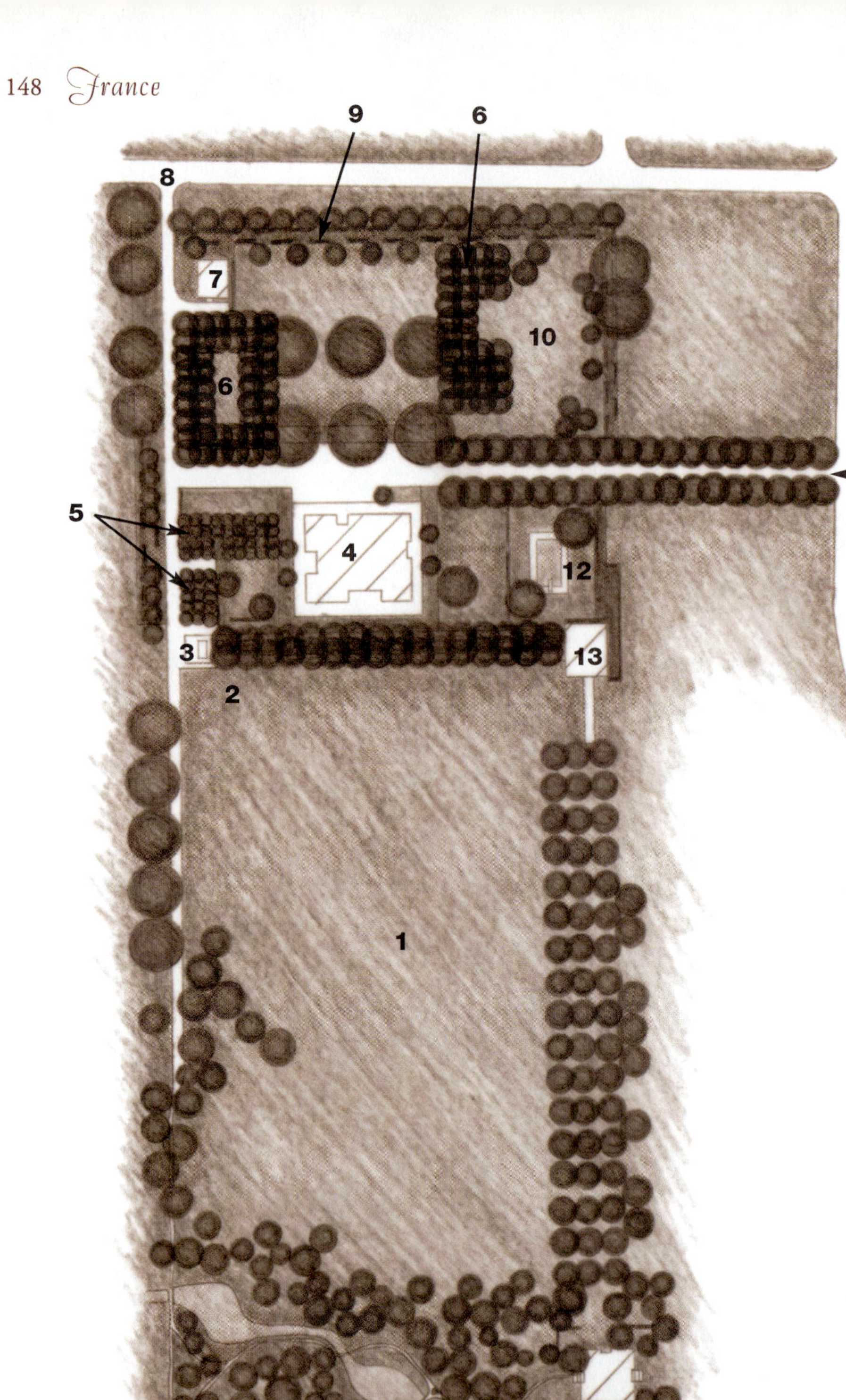

The Miller Garden

1. Meadow
2. Locust Allée
3. Sculpture
4. House
5. Groves of Flowering Trees
6. Apple Orchard
7. Greenhouse
8. Service Drive
9. Arborvitae Hedge
10. Playcourt
11. Main Drive
12. Swimming Pool Area
13. Fountain and Sculpture

The Miller Garden
Columbus, Indiana

The Miller Garden, a private residential garden in Columbus, Indiana, is a major work of twentieth-century modernist design in America. It also is a fine example of the influence of historic precedents in contemporary work. Built during the 1950s, the garden has gained national notoriety as one of the finest accomplishments of mid-twentieth-century modernist design. It became a paradigm for the design of modern landscapes of that period. Its designer, Dan Kiley, a nationally recognized landscape architect from Charlotte, Vermont, has described it as a "complete geometry on the land" (Hilderbrand 1999), undeniably influenced by the principles of French classicism. Ironically, Kiley, early in his career, rebelled against classicism, establishing himself as a modernist in search of a new design style that denied former traditions. But after visiting the gardens of France, and especially viewing the work of André Le Nôtre, French classicism became one of his main inspirations. Le Nôtre was an expert at organizing space on a grand scale. His gardens united architecture, garden, and site in a unique and masterful way. It is not difficult to see how his work influenced Kiley's design approach. Like Le Nôtre, Kiley uses natural elements such as hedges, orchards, *allées,* and rows of trees to organize space for both function and aesthetically pleasing composition (Figure 9–1). The design of the Miller

FIGURE 9–1 Blocks of ground cover, hedges, and rows of trees are used to organize space in the garden. Courtesy of Miller Gardens.

FIGURE 9–2 One of the main goals for the designers was to create a smooth transition between interior and exterior space. Courtesy of Miller Gardens.

FIGURE 9–3 A grove of flowering crab apple trees casts shadows on the ground in the "Adult Garden." Courtesy of Miller Gardens.

FIGURE 9–4 A strong linear path connects the house and the swimming pool area. Courtesy of Miller Gardens.

House, by architect Eero Saarinen, informed Kiley's overall planning of the garden. One of the main goals for both designers was to create an invisible transition between architecture and landscape, a spatial continuity between inside and outside. Kiley's solution was to create a series of garden rooms extending the interior plan out into the landscape without interruption (Figure 9–2). In his scheme, strictly ordered spaces transition into open, more relaxed areas of lawn and organized groves (Figure 9–3). Each area or garden room has a specific program: apple orchard, children's garden, entertainment or recreation space—yet all are united by a logical order of spatial progression. Functional relationships create the basis for his design. Spaces are logically proportioned and meticulously organized. Kiley's scheme leads one from space to space in a process of discovery (Figure 9–4). Each area is unique and inviting, informed by, but not attempting to copy historic traditions.

Chapter opening image courtesy of Miller Gardens.

FIGURE 9–5 A cross-axial view from the honey locust allée *to the open landscape below. Courtesy of Miller Gardens.*

Functional concerns are addressed with classic solutions designed in modern ways. For example, along the western edge of the garden terrace, adjacent to the house, Kiley arranged an *allée* of honey locusts (*Gleditsia triacanthos* "moraine") to shield the living area of the house from the afternoon sun (Figure 9–5). But beyond this functional solution, the double row of trees draws attention to sculptures placed on either end of its path. The spacing of the trees and the manipulation of the adjacent landform on the meadow edge of the terrace encourage cross-axial views from the *allée* to the vast, open landscape below. Traditionally, *allées* were created to direct views and movement down an axis it helped define, but here lateral views across the double row of trees are more pronounced than those down its central path. Its location also denies tradition. Rather than extending from the house or other spatial structure, the *allée* of trees begins and ends with sculptural art pieces and runs alongside the house, functioning more like an architectural element, uniquely creating a transition between built form and landscape, and setting up controlled or framed views to the west across the vast floodplain.

In the apple orchard, which appears to be spaced according to conventional rules, Kiley introduces a new twist by leaving a rectangular void at its center. When the open, rectangular area captures sunlight, one can begin to appreciate the gridded composition from an entirely new perspective. The rectangular shape of the entire arrangement is emphasized by this central

FIGURE 9–6 Shadow, color, and fragrance in the orchard accentuate the beauty of nature. Courtesy of Miller Gardens.

FIGURE 9–7 The stepped arborvitae hedge eliminates the "walled" effect characteristic of most hedges. Courtesy of Miller Gardens.

lighted space, and the static grid is transformed into a dynamic space as light moves through it (Figure 9–6).

Another rather ordinary landscape element given a new definition in Kiley's garden is the privacy hedge. In many gardens, hedges are functional but rather uninteresting as a planted form. In the Miller Garden, hedges are arranged in a unique way along the property's boundaries. Arborvitae *(Thuja occidentalis)* are grown in interrupted lines with sections stepped back to create less of a "walled" effect (Figure 9–7). The hedge is no longer seen simply as a border but as an interesting planted form that adds to the modern features in the garden. This same pattern is repeated in the design of an architectural screen constructed between the main entrance to the house and the car park (Figure 9–8).

FIGURE 9–8 The alternating rhythm of the stepped-back hedging at the property boundary is repeated in the architectural screen at the car park. Courtesy of Miller Gardens.

Kiley is not controlled by classic principles in his work, but his knowledge of the fundamentals of the French style and his appreciation for the lessons of spatial order and structure, especially in the work of Le Nôtre, are evident in his designs. The success of the Miller Garden derives from a blend of classic principles with a modern variation. The basic components of the garden are not changed, but their application takes on a new meaning. Planted forms are skillfully translated into a contemporary garden language that responds to the intended design program and appeals to a modern awareness. The Miller Garden is a lesson in the value of historic work to contemporary design.

Chapter 10 Landscape Expression: French Parks of Royal Heritage

Public gardens did not exist in Paris until modern times, but several royal gardens owned by kings and other dignitaries were open to the public. The owners maintained their privacy by enclosing part of the garden with walls or hedges and reserving that space for their own use. The principal gardens that were open to the public included the Jardin des Tuileries, the Jardin du Luxembourg, and the Jardin des Plantes.

Jardin des Tuileries

PARIS, FRANCE

Lying along the north bank of the Seine, in the heart of Paris, is one of the finest parks in France, a vast landscape rich in history, dating from the mid-sixteenth century. This park, once associated with a royal French palace, is the Tuileries. The palace no longer exists, having burned down in 1871, shortly after the siege of Paris. However, the gardens that surrounded it survive today, as one of the most frequently visited parks in Paris (Figures 10–1, 10–2).

Its royal history dates back to 1518, when Louise of Savoy, mother of François I (1494–1547), acquired the property. The origin of the name

FIGURE 10–1 A pleached allée *and perennial border frame meticulously groomed lawn areas at Tuileries.*

FIGURE 10–2 Park visitors gather around a circular pool.

Jardin des Tuileries Time Line

1518	Louise of Savoy, mother of François I, acquires the Tuileries.
1559	Queen Catherine dei Medici, wife of Henri II, adds to the property of Tuileries.
1564	Construction begins on the Tuileries palace and gardens.
1590	The Tuileries is ravaged during the civil war.
1594	Henri IV settles in Paris after the civil war and begins restoring the Tuileries palace. The gardens are redesigned by Claude Mollet.
1659–1672	Louis XIV commissions Louis Le Vau and François d'Orbay to make improvements to the palace and André Le Nôtre, the royal gardener, to redesign the gardens.
1715	King Louis XV, age five, takes up residence at the Tuileries.
1789	King Louis XVI is forced into Paris by the revolutionaries and makes Tuileries his chief residence.
1792	The palace is taken over by the republic. Improvements are made to the palace, and the gardens are redesigned in the popular English fashion.
1800	Napoleon I moves into the Tuileries. The palace is renovated, and the Triumphal Arch is built to pay homage to Napoleon's victory in the Austerlitz campaign.
1831	Louis-Phillipe takes up residence at the Tuileries and encloses part of the garden as a private space.
1852	The Orangery is constructed.
1859	Napoleon III expands the private garden.
1861	The Jeu de Paume is constructed.
1870	The empire collapses. The Tuileries is taken over by Prussian troops.
1871	Revolutionary government units set fire to the Tuileries palace, completely destroying it.
1883	The palace is demolished.
1889	The gardens are extended to the former site of the palace.
1900–present	During the 1990s, the gardens were revised by Louis Benech, Pascal Cribier, and Jacques Wirtz. Today, public ceremonies and other festivities are held in the park. The tradition of sculpture in the garden is continued, with several pieces installed throughout the twentieth century.

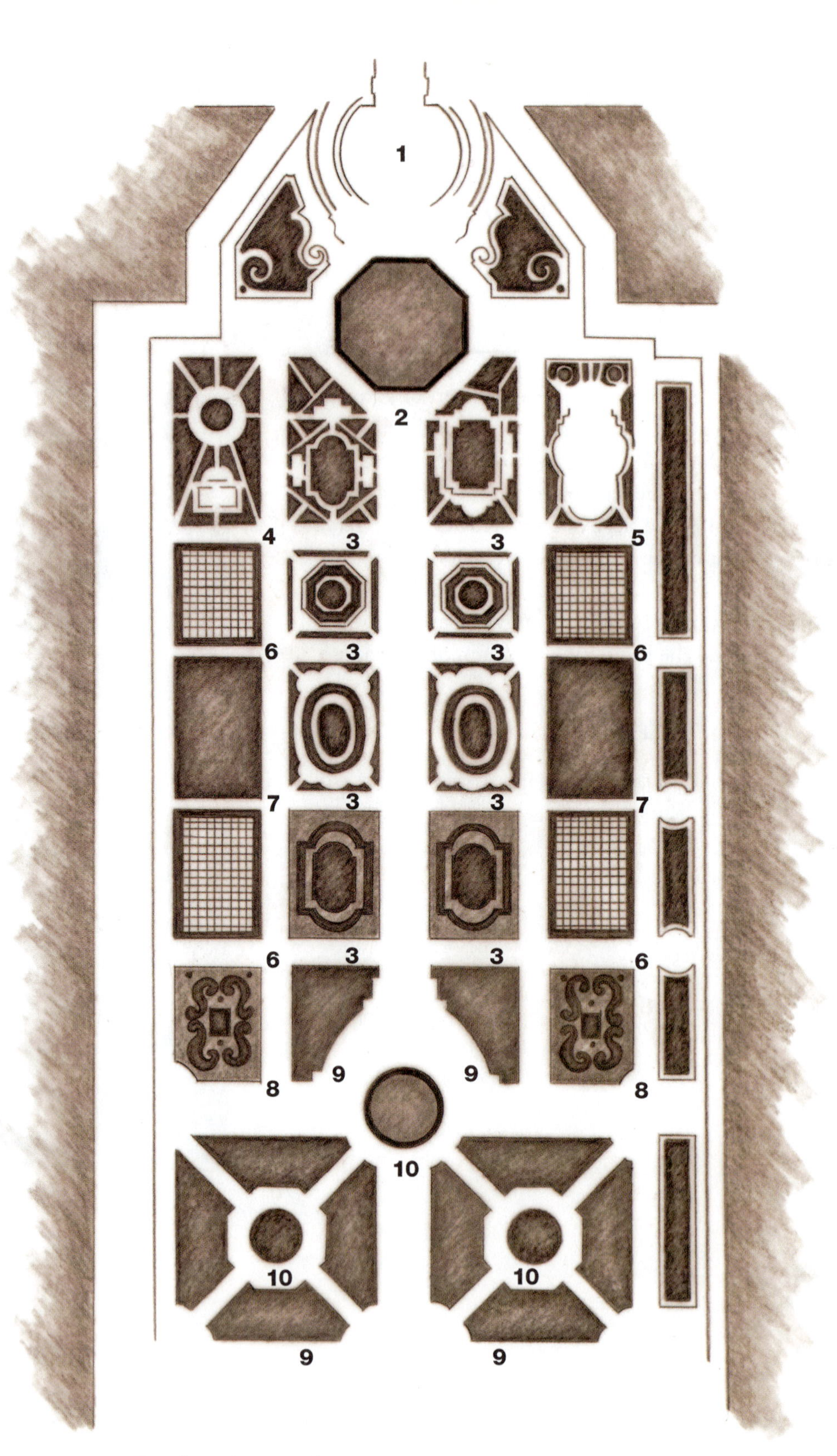

Jardin des Tuileries

1. The Horseshoe
2. Octagon Pool
3. Salles ou bassins de Gazon
4. Allées Plantes de Maroniers d'Inde
5. Salle de la Comedie
6. Petit bois plantes
7. Bosquets
8. Parterre de Gazon
9. Parterres de Broiderie
10. Les 3 bassins bordez de Gazon

Tuileries is traced to an old tile factory (*tuileries* is French for "tile works") that occupied the site up until the sixteenth century. The property was later passed to Henri II (1519–1559). After his death, in July 1559, his wife, Queen Catherine dei Medici (1519–1589), began to acquire property adjacent to the site as part of her plan to construct a palace there. In 1564, construction began on the new palace designed by architect Philibert Delorme, who also worked on the royal chateau at Fontainebleau. Delorme died in 1570, and architect Jean Bullant completed the work. For the gardens, Catherine, fond of the Italian style she knew in her youth, commissioned Bernard de Carnessequi from Florence to construct an Italian garden complete with fountains, a maze, and a rustic grotto. Kitchen gardens, orchards, and planting beds rich with flowers were added, making it the most impressive garden in Paris and a place where lavish festivities were staged.

Henri IV (1553–1610) ascended the throne in 1589, but the civil war prevented him from entering Paris and occupying the Tuileries right away. The king finally settled in Paris in 1594 and began making improvements to the palace as well as to the gardens, though he never actually lived there. One of his most significant ideas was the proposed connection between the Louvre and the Tuileries through the construction of a "Long Galerie." Henri also made contributions to the landscape surrounding the palace, employing garden designer Claude Mollet to restore the gardens, which had been substantially destroyed by the civil war. Mollet's scheme included a handsome arbor constructed over a path extending the entire length of the garden. Beside it, a row of Mulberry trees was planted to nourish the silkworms that were being raised by Henri in an effort to spawn a new silk industry in France. Though the king never made Tuileries his home, the gardens were one of his favorite places to spend his leisure time.

When Louis XIII became king in 1610, he was only a nine-year-old child. The gardens at Tuileries were more of a place of amusement for the young boy than anything else. There he could hunt, practice shooting, and ride horses. Louis XIII's niece, the Duchesse de Montpensier, was given the property as a gift from the king when she was born in 1627, and she resided there as an adult up until the Fronde (a rebellion led by the Parlement of Paris from 1648 to 1652, against the taxation policies of the crown). It was then that Louis XIV had her removed from the palace for her suspected participation in the rebellion. After the Fronde, an effort was made to establish a royal presence in the city. Work immediately began at the Tuileries to establish a respectable royal residence. Architects Louis Le Vau and François d'Orbay accomplished the refurbishing of the palace between 1659 and 1666.

The work on the palace was accompanied by a complete reworking of the gardens during that same period. Louis XIV's finance minister, Jean-Baptiste Colbert (1619–1683), commissioned landscape designer André Le Nôtre to create a new scheme that would accommodate the newly constructed palace. Le Nôtre significantly altered the original gardens of Catherine dei Medici with a plan that took nearly six years to implement. He removed the existing road between the palace and the gardens, incorporating a terrace in its place. From the terrace, a pleasing view of newly designed *parterre* gardens could be enjoyed. These gardens were meticulously crafted into ornate patterns from tightly clipped evergreen shrubs and filled with an assortment of flowers. The composition was organized around three circular water basins, a pair of equivalent size on the north and south sides of the central path and a slightly larger pool centered on it. The central path, or Broad Walk, slightly over 1,000 feet in length, began at the central pond and was flanked by parallel walks on either side. Secondary paths running perpendicular to the main avenues divided the garden into several smaller sections, most of which were planted with their own distinctive scheme. Two terraces descended from the garden level to the Seine. The one to the south was planted with trees that framed several beautiful views of the river, and the one on the north side of the terrace overlooked the flower gardens. The central path ended at a large, octagonal-shaped pool spanning nearly 200 feet in diameter. The axis was later extended beyond the park, out to what would become the famous Champ-Elysées. Most of the garden construction was completed by 1672.

It was during the reign of Louis XV (1710–1774) that the garden became more completely integrated with the city. The garden's axis to the west toward the Champ-Elysées was made continuous by building a bridge over a hollow that had previously broken the path. On the west, the Place de la Concorde was created to provide a grand exit from the park. The king made Versailles his chief residence and used the Tuileries as a sort of annex to house dignitaries. It also functioned as a cultural center, where theater performances were staged and concerts were held. The gardens functioned as a stage for grand celebrations and as a place where the aristocratic culture could gather to socialize.

The royal family continued to live at Versailles through the reign of Louis XVI. It was not until the start of the French Revolution in 1789 that the king would occupy the Tuileries, being forced into Paris by the revolutionaries. Three years later, the palace was invaded and the monarchy ended, marking the beginning of the new republic. The palace became the seat of the new government, and both it and the gardens underwent a substantial transformation. The interior of the palace was embellished by decorations confiscated from other royal residences. In the gardens, the groves were redesigned in the popular English fashion. *Parterres* were replaced by an open lawn, and various tree species were planted in informal groupings, invoking a more natural-looking style. Statues were placed in the park, with a host of sculptures from different periods displayed around the grounds, creating an open-air museum, a tradition that continues today. By the end of the century, the park had become a popular public venue, one of the few open green spaces available to people living in the city.

FIGURE 10–3 Napoleon's Triumphal Arch.

On February 19, 1800, Napoleon, soon to become emperor, moved from the Palais du Luxembourg to the Tuileries. The palace was renovated to accommodate him and his imperial court, and Napoleon's Triumphal Arch was constructed, creating a grand entrance (Figure 10–3). This arch, inspired by the arch of Septimus Severus in Rome, recalled one of the greatest battles won by Napoleon, the Austerlitz campaign of 1805. It is the work of architects Charles Percier and Pierre François Léonard Fontaine. The monarchy was reinstated with Louis XVIII, who was followed by Charles X. The gardens underwent little change during those years, aside from some restoration work.

When Louis-Philippe came to power in 1830, he had an area of the garden made private by surrounding it with a ditch. He embellished this "reserved space" with fine sculptures and had his architect, Fontaine, create two new, large flowerbeds. In 1848, Louis-Philippe was overthrown during the Second Republic and forced to flee Paris. Napoleon III came to power with the return of the empire, beginning in 1852. He chose to reside at the Tuileries and made it and the adjoining Louvre the center of government. The gardens were further embellished with the planting of rare and exotic species, and the reserved or private garden was extended west to the crosswise path. On the west end of the garden, two new buildings were constructed, the Orangery and the Jeu de Paume.

In 1870, the empire collapsed, and the Tuileries was taken over by Prussian troops. The palace remained empty through the next year, with the official government of the Third Republic set up in Versailles. In May 1871, revolutionary government units set fire to the palace and the Louvre library, completely destroying it. The building remained as a ruin until 1883, when Monsieur Achille Picard ordered that it be demolished. Two years later, gardens were developed on the site of the former palace, and new sculptures were added to the garden.

Throughout the twentieth century, the park hosted numerous festivities, from car shows to government events. Its central location in Paris makes it easily accessible to a great number of people. Various forms of entertainment and activities in the park cater to young visitors. Carts and kiosks selling ice cream and toys can be found along the garden paths, and many a child can be spotted floating model boats in any one of the ponds.

The Tuileries' cultural role has asserted itself by establishing the Orangery and the Jeu de Paume as exhibition galleries for various art installations, both local and foreign. The tradition of sculpture in the park has continued, with the inclusion of twentieth-century work by artists such as Jean Dubuffet, Aristide Maillol, and Henry Moore.

Jardin du Luxembourg

1. Palais du Luxembourg
2. Fontaine de Médicis
3. Octagonal Pool
4. Tennis Courts
5. Children's Playground
6. Orangerie

Jardin du Luxembourg

Paris, France

The history of the Jardin du Luxembourg begins with Marie dei Médicis (1573–1642), who arrived in Paris in 1601, a few months after her marriage to King Henri IV of France. She took up residence at the Louvre but went on to acquire other estates, both before and after the death of the king in 1610. The Palais du Luxembourg was one of the many estates she acquired. As queen regent during the minority of her son, the future King Louis XIII (1601–1643), Marie purchased a house from the duke of Luxembourg in order to lodge visiting foreign princes and ambassadors. What followed was the creation of an extraordinary palace and gardens, which remained a royal possession until the French Revolution (Figure 10–4).

It seems that Marie dei Médicis likely purchased the Luxembourg property with the intent of creating a grand residence to accommodate her own future living needs rather than simply to house visiting dignitaries. She was certainly aware that her son, the young king, would ultimately occupy the Louvre with his wife, which would require her to take up residence elsewhere. Thus it was in the queen's best interest to create a suitable palace that she could move into once removed from her royal apartment at the Louvre.

Marie purchased several other properties adjacent to the Luxembourg property, and in 1611, she commissioned architect Salomon de Brosse (1565–1626) to develop a plan for a new residence and garden there.

FIGURE 10–4 Park visitors gather in front of the former Luxembourg Palace, now the home of the senate.

Jardin du Luxembourg Time Line

1600	Marie dei Médicis and King Henri IV of France marry.
1601	Marie dei Médicis arrives in Paris and takes up residence at the Louvre.
1610	King Henri IV dies.
1611	Marie de Médicis purchases the home of the duke of Luxembourg and commissions architect Salomon de Brosse to develop plans for a new palace.
1612	The queen hires Nicolas Descamps to create a plan for the palace's gardens.
1615	Construction begins on the palace. Work continues through the next decade.
1642	Marie dei Médicis dies.
1791	The Convention turns Luxembourg into a prison during the republic.
1801	Napoleon makes the former royal palace home to the senate.
1800s	The gardens are redesigned and established as a public park.
20th c.	The park attracts visitors by incorporating many recreational activities, such as basketball, bocce ball, and tennis, as well as a large playground for children.

Fond childhood memories of growing up as the daughter of the grand duke of Tuscany in the Pitti Palace of Florence, Italy, instilled in her a desire to develop a similar plan for her new palace. She had de Brosse consult the architectural drawings of the Florentine palace, though he claimed he received no inspiration from them. Instead, he credited his Uncle Androuet du Cerceau (1544–1602) as the source of his creativity, particularly his design for the chateau of Verneuil, a French creation but with many Italian-inspired features. De Brosse's design for the Luxembourg palace is rather typical of French chateaus built during this period. The plan consists of a large central block, crowned by a cupola and joined by two wings that connect to two pavilions on either side. Unique to de Brosse's work is the articulation of exterior detail and ornament that one can safely say is owed to Renaissance examples. Construction of the palace began in the spring of 1615 and progressed slowly because of political and economic hardships. A document written in 1623 reported that the palace was still only partially completed nearly a decade after construction had begun.

In 1612, the queen hired Nicolas Descamps to create a plan for the palace's gardens. His work displayed some Italian influences in both organization and design techniques. The main garden consisted of *parterres* organized on either side of a central axis. The termination of the *parterre* garden in a semicircular form was further evidence of Italian precedents, as many Italian garden

FIGURE 10–5 Areas of entertainment are nestled within shady groves of trees.

designers had experimented with this technique. However, many elements of the design anticipated a style that could be regarded as distinctly French—numerous *allées* of chestnut and linden were planted to identify and shade walkways, and highly decorative *parterres* were created with clipped boxwood and yew. Color in the garden was reserved for the Jardin Fleuriste (or "Flower Garden"), laid out below the palace windows. This garden was planted with a multitude of brightly colored annuals arranged in beds for the queen's delight.

Luxembourg remained a royal palace until the revolution, at which time the National Convention turned it into a prison. In 1801, Napoleon made it the home of the senate, which it remains today. During the nineteenth century, much of the garden's seventeenth-century layout changed when it was redesigned and established as a public park. Many forms of entertainment were incorporated into the new layout, accommodating visitors of all ages. Today, areas for tennis, basketball, and bocce ball are carved out from a tight grid of shade trees (Figure 10–5). A large playground with brightly colored equipment has become a favorite of the children who visit, although many other unique and fun activities in the park cater to young Parisians, including pony rides, a merry-go-round, and a puppet theater (Figure 10–6). The park, known for its lighthearted and festive atmosphere, is a popular attraction for families with small children, though on any given day it is possible, and even likely, that a visitor will be challenged to a game of bocce ball or a round of tennis.

FIGURE 10–6 Activities in the park for children include a playground filled with brightly colored play equipment.

Jardin des Plantes

1. Grande Galerie
2. Grande Galerie Courtyard
3. Galerie de Minéralogie et de Géologie
4. Galerie de Botanique
5. Menagerie
6. Garden Plots
7. Place Valhubert
8. Seine

Jardin des Plantes

PARIS, FRANCE

On the Left Bank of the Seine, in the fifth arrondissement, are 70 acres of walled garden and a group of research and teaching institutes dedicated to the natural sciences. This campus-like park, one of the first of its kind, is called the "Jardin des Plantes." From its earliest history, it fulfilled the role of both a public garden and a site for botanical research.

In 1626, Louis XIII, at the urging of his personal physician, Guy de la Brosse, authorized the construction of the garden. It was called the "Jardin Royal des Plantes Médicinales" (the "Royal Garden of Medicinal Plants") and was created by de la Brosse to meet the needs for herbs being used for medicinal purposes. For over a century the garden primarily served a utilitarian role, though it continued to develop as something more than a medicinal garden. It emerged as a leading research center, which included many of the sciences of nature, such as biology, entomology, zoology, and, of course, its earliest discipline, botany.

The garden occupies an area in the Quartier de Saint Victor, southeast of the center of Paris. It essentially lies along the riverbank in an area of the city populated with students and faculty from the neighboring Faculté des Sciences de l'Université de Paris (Faculty of the Sciences of the University of Paris). The main entrance to the garden opens onto a wide promenade, beyond which is a *parterre* garden. The *parterre* is bordered on either side by double *allées* of **pleached** (tightly clipped and intertwined branches) plane and chestnut trees (Figure 10–7). The planting beds are filled with a brilliant display of colorful bedding plants, including cannas, chrysanthemums, and dahlias, and a broad range of summer annuals (Figure 10–8 and Figure 10–9). On the east side of the gardens is the Galerie de Zoologie (Gallery of Zoology) and other buildings associated with research and teaching, including the Galerie de Minéralogie et de Géologie (Gallery of Mineralogy and Geology) and the Galerie de Botanique (Gallery of Botany).

In 1739, French naturalist George-Louis Leclerc de Buffon became the director of the garden. Under his direction the botanical collection was expanded to include plants other than medicinal herbs. He enlarged the garden, nearly doubling its size, and added many exotic specimens collected from various plant expeditions that he assigned to the far reaches of Asia, Africa, and the Americas. The early introduction of exotic and often tender plantings led to the construction of heated glass buildings, later known as "greenhouses." A botanist, Sebastian Vaillant, working for King Louis XIV's personal physician, Guy Fagon, was essentially the progenitor of the greenhouse concept in the French garden. Vaillant created a rather crude miniature glass structure in

FIGURE 10–7 An allée *of pleached plane trees borders the park.*

Jardin des Plantes Time Line

1623	Louis XIII acquires the property for the Jardin des Plantes.
1630–35	The Jardin Royal des Plantes Médicinales (the "Royal Garden of Medicinal Plants") is created by the king's personal physician, Guy de la Brosse, in order to satisfy a need for herbs used for medicinal purposes.
1739	French naturalist George-Louis Leclerc de Buffon becomes director of the garden. Buffon expands the botanical collection to include plants other than medicinal herbs. Glasshouses are constructed to house exotic specimens.
1793	A small zoo of approximately 14 acres is added to the garden.
20th c.	The garden gains in popularity as a public park and continues its dual role as a place of leisure and academic research.

FIGURE 10–8 The planting beds are filled with a mixed collection of flowers.

FIGURE 10–9

order to accommodate a tender specimen that he had acquired. His model was followed by larger, more sophisticated versions made of iron and glass during the nineteenth century. Today, many tropical plants are grown under glass in the gardens, including a collection of over 1,000 species of orchid.

The remains of what was once the École de Botanique ("school" of botany) are on the north end of the garden. The site of the original gardens, this small, cultivated plot, is where the original collection of herbal plantings was grown and studied. Today, nearly 4,000 species are planted in these research gardens, especially some of the more recent varieties of herbs with health-giving or medicinal properties.

In 1793, following the French Revolution, the National Convention (elected deputies of the revolutionary government) added a small zoo of approximately 14 acres—the ménagerie, at the northwest corner of the gardens.

It is the oldest public zoo in the world. The idea of a ménagerie was not a new one, however, until then, such entertainment was limited to royalty and nobles. In fact, many of the animals that were brought to the ménagerie were collected from Versailles, which at that time had been taken over by the republic. The ménagerie continued to acquire many exotic animals, including giraffes, lions, jaguars, and bears.

The garden continues its dual role as a place of leisure and academic research, consistent with the program established in its earliest days. It is by no means a famous place, nor is it a popular destination for most tourists. It is noticeably a local public garden, distinctly inhabited by Parisians and seemingly very popular among them.

Part Three

England

Environment

England occupies the region of Great Britain just east of Wales and south of Scotland. It is narrowly separated from the continent of Europe by the English Channel. The country has one of the most extraordinary landscapes for its size in all the world. It is characterized by a diverse physiography and substantial climatic differences from north to south.

England's terrain is divided into two distinct zones, the highlands and the lowlands. The dividing line can be drawn from the southwest to the northeast corners of the country, with the highland region occupying the north and the lowland territories lying to the south. The principal highland region to the north includes the Pennine chain of mountains. The greater part of the area to the west of the northern range of these mountains is referred to as the "Lake District," known for its picturesque beauty. A landscape consisting of several lakes, or meres, and a number of winding streams, it is indeed one of the most beautiful regions of England. Low hills and rolling plains occupy most of the lowlands of central and eastern England. The region to the west, from the southern portion of the Pennine hills to the Bristol Channel, is known as the "Midlands." This area includes the Industrial West Midlands, sometimes called the "Black Country" because of the intensive industrial development that occurred there during the nineteenth century. Not far from these industrial communities is an area known for its beautiful farm villages and charming stone houses made from Cotswold stone. This region, called the "Cotswolds," is known for its successful farm communities that benefit from fertile land and an agriculturally friendly climate. The Cotswolds region also is renowned for its many historically significant gardens and landscapes. To the east is the Fens, or the Fenland, a vast, artificially drained marshland. Between 1637 and 1987 most of the Fenland, once

occupied by swamps and marshland vegetation, was drained under the direction of a Dutch engineer hired by the earl of Bedford. Today, less than 1 percent of the marsh remains.

England enjoys a moderate climate with few instances of either extremely cold or hot temperatures because of the Gulf Stream, a warm current of the Atlantic Ocean that has a warming effect on the waters that surround England. This effect, in combination with cooler air temperatures, creates moisture and warmer ambient temperatures in the winter. It also produces cooler temperatures during the summer months. Climatically, England has the greatest variation in temperature and rainfall from east to west, with the warmer temperatures and greater precipitation occurring to the south and the cooler temperatures to the north and east. The mean annual temperature in the south is 52°F, while the north averages 48°. The average rainfall amount is about 30 inches, with October being the wettest month. The fog and mist that England has long been known for are indeed common, especially in the Pennine and inland areas, where it rises above the verdant hills and valleys. The moist atmosphere combines with cooler temperatures, resulting in especially robust plantings and lush green lawns.

Social and Cultural History

The story of modern English history begins in 1485 with the accession of Henry VII (1485–1509), the first of the Tudor kings. Prior to his rule, the country had been devastated by war and epidemics. The government was weak at best, and law and order were nearly absent from society. England was in a desperate situation, its population severely diminished and its economy in a state of ruin. It was Henry who, after restoring the central government, brought peace and prosperity back to England and law and justice to its societies. England continued to prosper during the sixteenth century, bolstered by overseas trade.

When Henry VII died in 1509, his son, Henry VIII, ascended the throne. Young Henry, bright and ambitious, inherited an economically sound government. He has long been praised for his accomplishments and for England's continued progress throughout his reign but is most remembered for his preoccupation with securing a male heir. Frustrated with his first wife's inability to provide him with a son, he decided to divorce her, and he asked for support from the Catholic Church. When the Church refused to endorse his request, he broke all ties with it. This led to what many Englanders call his most notable achievement, the break with the Church of Rome, which made England the first of many countries to be ruled under a Protestant Crown.

With this change, monasteries throughout England were destroyed, and their land was redistributed, mostly to the gentry classes, which strongly supported Henry and the new Protestant church. This activity essentially shifted the political, economic, and cultural power of England to the landholding gentry. The early history of English gardens and gardening during

the sixteenth and seventeenth centuries is generally associated with these men and with the royals. There was a correlation between disposable wealth and the extravagance of the garden. The royals and the affluent were able to create and maintain magnificent gardens, employing the most skilled garden designers and caretakers.

Gardens were formally arranged and often walled or enclosed by hedges, especially near houses. This offered some privacy to the owners, especially to those who made their garden available to others. These enclosed areas also protected vegetables and flowers planted near houses from destruction by wandering animals. Gardens typically evolved as a series of individual spaces, each with its own particular function—growing fruit, herbs, vegetables, or flowers.

Later, Italian works inspired more ideas for garden design, as the relationship between Italy and England improved. Few Englanders actually traveled to Italy, but ideas were spread through available publications and also through the experience of art shared by those who had traveled there and acquired paintings and sculptures. There was little garden making during the mid-to-late 1600s, as wars hindered England's economy and spirit. Land was taxed heavily to pay for the war efforts, leaving large landholders less disposable income for creating gardens.

The early eighteenth century was a time of peace and the beginning of a new governing party called the "Whigs." Leaders of this group, eager to

firmly establish themselves and share their political philosophies, created gardens with political undertones, designed as statements of their party's values. These gardens displayed a vast amount of symbolism in the form of sculpture and buildings that conveyed Whig politics and beliefs. England continued to prosper under Whig rule, and individuals' lifestyles improved. People were healthier, wealthier, and more educated.

As Quest-Ritson points out, the classics were the basis of education, and a prerequisite for good taste. It was this knowledge of the ancient world that elevated individuals' status, setting them apart from the "common" society. The wealthy classes were not only classically educated but also worldly and refined, with a broad exposure to books, art, and other culturally acceptable interests, including gardening. It was thought that these activities developed good taste, and taste was the mark of high society. It was the defining characteristic of a "gentleman" (Quest-Ritson 2001).

A classical education and garden making went hand in hand, and as the eighteenth century evolved, classical fashion appeared in many English gardens. Classical sculpture, temples and ancient buildings were incorporated into the design of gardens as physical representations of classical writings and paintings. Classically inspired gardens were designed to demonstrate one's wealth and social position. They were to be experienced and enjoyed only by a cultivated society that understood them, as the uneducated had neither the wisdom nor the taste to appreciate their beauty and significance.

Throughout the eighteenth century, there was a growing desire among landowners to improve their properties by increasing their capital value. Landholdings were consolidated for more efficient management and to build up a position in one or more areas in order to gain political control of local government. Parliamentary enclosures helped the large landowners acquire additional acreage by allowing the transfer of commonly held pasture land, or unoccupied "waste" fields, to private ownership. This contributed to the rise of the landlord class, whose economic success, resulting from improved farming methods, marked the beginnings of commercial agriculture as we know it.

and successful. Many fine examples of these country estates with their naturalistic gardens have been thoughtfully preserved throughout England, where they survive as a living legacy of a popular and beautiful English tradition. Since the late 1940s, many of these have been made available to the public through the efforts of the National Trust—a government organization dedicated to the preservation of historic properties and gardens throughout England.

Chapter 11

Elements of English Garden Design

The English gardening style often is described in terms of the eighteenth-century English landscape garden, an original approach to garden design that was perhaps Britain's most important contribution to garden history. But even before this revolution in gardening, England had established a tradition of garden making dating back to the Middle Ages. The early English gardens were small, private enclosures filled with flowers, herbs, and vegetables neatly arranged inside of garden walls or hedges. Most gardens were practical, and few were actually designed.

Not until the early sixteenth century, under the reign of the Tudors, did the garden show signs of artistic influence. Major progress in garden development occurred during the reign of King Henry VIII. His ties to Italy, though short-lived, made him familiar with popular Renaissance customs. Hampton Court, one of the first royal palaces, shows signs of Italian influence as well as elements of French and Dutch design. In the original gardens developed by Henry, spaces that would earlier have been joined without any attention to spatial sequence or artistic detail were instead logically organized and integrated following the principles of classic Italian design. Various garden features also were inspired by earlier gardening techniques. For example, Queen Elizabeth's favorite, the **English knot garden** (a garden in which low-growing evergreen shrubs or sometimes herbs are planted in an intertwining pattern resembling a knotted rope, with the spaces in between typically filled with gravel or flowers, (Figure 11–1 and Figure 11–2) and bears resemblance to

FIGURE 11–1 An English knot garden. Courtesy of Dr. Michael N. Dana.

FIGURE 11–2 An English knot garden. Courtesy of Dr. Michael N. Dana.

the Italian parterre. The French influence was introduced to the English garden by Charles II, who returned to England in 1660 after being exiled to France. It was not long after his return that he began to display his fondness for French design by incorporating aspects of the French garden style into the gardens that he created at Hampton Court. Grand avenues, axial vistas, elegant fountains, and ornate *parterres* became components of his new garden and soon found their way into other new garden developments across England. At Chatsworth House in Derbyshire, gardeners George London and Henry Wise also developed a garden plan with formal patterns and classically inspired garden features.

By the eighteenth century, a new, informal style emerged from a growing discontent with the formality of gardens informed by Italian and French models. This new landscape approach was inspired instead by literature and classical paintings. Authors such as Joseph Addison, Alexander Pope, and Stephen Switzer wrote persuasively about the virtues of embracing nature, and they described the beauty inherent in natural landscapes. Picturesque

Figure 11–3 Monuments, temples, and sculpture in early-eighteenth-century gardens gave expression to the thoughts and beliefs of garden owners.

scenes of idyllic landscapes painted by seventeenth-century artists such as Claude Lorrain and Nicolas Poussin provided further inspiration and were even copied, becoming real designs created for wealthy landowners. Gardens were designed as a series of staged scenes uniquely blending nature and art. Monuments, temples, and sculptures, imbued with classic symbolism, gave expression to the thoughts and beliefs of the gardens' owners (Figure 11–3).

Charles Bridgeman and William Kent were two popular designers during this period. Each made notable contributions to this new and original style by creating gardens that became influential models. Bridgeman was one of the first to attempt to create a more natural, informal garden. His most popular work, and perhaps his greatest achievement, was the garden at Stowe in Buckinghamshire. Though not responsible for the entire landscape (much was completed or revised after his death), Bridgeman introduced the original layout of the garden. His successor at Stowe was William Kent, who altered some of Bridgeman's work and created new gardens that were even more informal. The "natural" landscape style continued to gain in popularity. Nature was no longer something to control. Instead, it was copied and celebrated for its "wildness." Water, topography, and plantings were altered and arranged to create a landscape that blurred the boundaries between the "natural" and "designed." A unique invention called the "**ha-ha**" (a walled ditch separating the garden or park from fields of grazing livestock; Figure 11–4) further helped establish that visual continuity, making the landscape appear to extend without limits to the neighboring countryside.

FIGURE 11–4 A ha-ha designed into the landscape at Rousham Gardens in Oxfordshire separates the garden from an adjacent field.

Other garden successes in the developing English landscape style include Rousham in Oxfordshire, also designed by William Kent, and Stourhead in Wiltshire, a popular garden created by wealthy banker Henry Hoare. Both gardens emphasize nature's ideal aspects and convey a spirit of antiquity through an itinerary of symbolic references.

The eighteenth-century English landscape progressed with an even greater movement toward "naturalness." Popular designer Lancelot "Capability" Brown was even more intent on creating a purely "natural-looking" landscape than his predecessors. His style avoided any sign of human intervention. Casual groupings of trees, curving lakes, and gentle, rolling hills and grazed meadows were characteristic of Brown's gardens (Figure 11–5). His creations were simple and informal, so close to true nature that many went unnoticed, mistaken for the natural countryside. Much of his work involved redesigning formal gardens in the popular new fashion. One of his finest accomplishments was Blenheim Palace in Oxfordshire. Brown's work there transformed an existing landscape by linking several well-designed, but disconnected, garden arrangements into a composition of distinctive garden features. Views in the garden were encouraged or denied to link various areas and to improve one's experience. Many significant features in the garden created by previous designers were enhanced and given new purpose by Brown's contributions.

English Victorian landscapes of the nineteenth century were diverse, with elements of many different styles—Italian, French, and others merged

FIGURE 11–5 Casual groupings of trees, gently rolling hills, and grazed meadows were characteristic of "Capability" Brown's signature landscape-design style.

in a dizzying display of ostentation. Gardens were being built for the new wealthy class of industry, a society that was not quite sure how to exploit their success. They built homes and gardens that were grand in scale and filled with extravagant details and decoration, but these new developments were characteristically lacking in good taste. By this time, design philosophies had changed, and the natural-looking landscapes made popular during the eighteenth century by Brown and others were no longer in fashion. Tastes trended toward classically inspired design. Elements of the Italian and French styles returned to gardens, but in a disorganized, almost whimsical way. *Parterres,* fountains, and sculptures, while beautifully designed

FIGURE 11–6 Many exotic plantings appeared in nineteenth-century gardens as a result of the popular interest among the wealthy for collecting and displaying foreign plants.

and impressive on their own, were united with other period elements and traditions in a rather tasteless and an often distractive semblance. Plant collecting was popular during the period, thus a wide variety of trees, shrubs, and flowers began to appear in the garden (Figure 11–6). Biddulph Grange in Staffordshire, developed in the first half of the nineteenth century, incorporates many exotic plantings as its garden's main feature. Plants are combined in a collection of simulated environments meant to replicate their native habitat.

The new technologies developed during the century allowed increased production of herbaceous plants, especially annual flowers, which became popular in gardens. Flowers were combined and arranged in fanciful patterns on the ground, a technique that became known as **carpet-bedding** (a nineteenth-century Victorian practice of arranging low-growing foliage plants of the same height in patterns that appeared carpetlike). Another influence affecting the design of Victorian gardens was the invention of the lawn mower, which gave lawns a new and desirable appearance, making them increasingly more popular for new garden developments.

By the mid-nineteenth century, there was a renewed interest in the organized nature of the seventeenth-century formal gardens. The garden was again seen as a place in which to live, and it was designed as a sequence of individual garden "rooms," separated by walls or hedges. But the question was—should the gardens be arranged formally or informally? Author William Robinson developed theories about the benefits of designing with nature, publishing his opinions in his widely read text *The English Flower Garden.* He was inspired by the beauty of plants in their natural surroundings and sought to transfer that same sense of beauty to the garden. He encouraged the use of self-seeding perennials to foster naturally evolving plant compositions filled with colorful blooms and exciting combinations of interesting forms and textures. These ideas evolved into a style that became popularly known as the English cottage garden style, inspired by the small and practical rural cottage gardens developed by the working-class society of the nineteenth century.

FIGURE 11–7 The late nineteenth- and early twentieth-century gardens were a combination of both formal and informal elements of design.

Robinson was not, however, unchallenged. Some, like architect and writer, Reginald Blomfield, disapproved of Robinson's ideas, arguing that the garden, as an extension of the house, should follow the same organizing principles of design to unite the two in a single composition. Blomfield, author of *The Formal Garden in England,* published in 1892, was Robinson's strongest opponent, arguing that visual order could only be achieved through "refinement and reserve" (Hobhouse 2002) and by creating an organized relationship between the house and the garden. These divergent theories of gardening were brought together in the work of architect Edwin Lutyens and landscape designer Gertrude Jekyll. Together they designed gardens that employed both the formal architectonic style and the informal, naturalistic approach (Figure 11–7). One of their finest garden achievements was Hestercombe in Somerset. There Jekyll proposed naturalistic plantings of various colors, textures, and forms as a way to soften architectural lines and features created by Lutyens. She used the garden's organized framework as a canvas upon which to create artistic compositions from a palette of hardy perennials, biennials, and annuals arranged in informal clumps and drifts.

Sissinghurst garden in Kent is another significant accomplishment in the gardening tradition encouraged by William Robinson and made popular by Lutyens and Jekyll. Organized as a series of outdoor rooms, each garden space was designed to accommodate some aspect of the owners' lifestyle. Bold plantings bursting with color and varied textures are displayed within a framework of old brick walls and tall evergreen hedges.

A new interest in this period in design was reborn in America in the 1980s and 1990s, when herbaceous planting became popular in residential

FIGURE 11–8 The informal or natural-looking garden style is still popular today.

gardens. The informal and natural characteristics of this freer style of gardening is chosen by those who view the "wildness" as an escape from the rigidity and tension of worldly pressures (Figure 11–8). There is a restorative spirit that arises from a closer association with nature, and that spirit is as strong today as it was at the dawn of the twentieth century.

Chapter 12

Landscape Expression: The Renaissance Garden in England

England's participation in the Renaissance movement was rather slow to develop. Prior to King Henry VIII's ascension to the throne in 1509, the country was preoccupied with several military campaigns that slowed the progress of new building. When Henry came to power, the country was in a peaceful state, united and economically strong. It was during this period that major development occurred on royal estates. Henry's ties to Rome, mainly through the Catholic Church, began strong and thus made him familiar with Renaissance progress in Italy. But that relationship soon deteriorated, resulting in a more or less impaired understanding of the Renaissance principles of design. Artisans from other countries employed in England's building projects applied classic Renaissance principles in their work, but with a compromised understanding that was less confident than their Italian contemporaries. During the sixteenth and seventeenth centuries, English gardeners became increasingly more competent in methods of Renaissance design. Their work combined original ideas with techniques borrowed from gardens created in other countries such as France and Holland.

The royal gardens at Hampton Court are a product of Italian, French, and Dutch influence. The contributions made to the gardens by a succession of kings and queens convey the progression of garden design in England from the early sixteenth century. Well preserved and meticulously maintained, these gardens are a valuable tool in the discovery of early English gardening style.

The chapter opening image depicts the garden adjacent to the east front of Hampton Court, drawn by Highmore and engraved by J. Tinney. (partial drawing)

Hampton Court
MIDDLESEX, ENGLAND

Hampton Court was one of the first great royal palaces and gardens created during the reign of King Henry VIII, beginning in the early 1500s. It developed as a symbol of the new monarchy's power and prestige and established a precedent for the future building of royal estates across England. Purchased in 1514 by Cardinal Thomas Wolsey (1471–1530), lord chancellor to Henry VIII, it was one of several properties owned by the wealthy cardinal, who valued it particularly for its splendid location on the river Thames (Figure 12–1).

In an effort to salvage a deteriorating relationship with his master, Wolsey offered his estate to Henry VIII in 1525. The king, perhaps envious of the cardinal's extravagant property, accepted the offer and made Hampton his chief residence. Wolsey, despite his generosity, eventually fell out of the king's favor, escaping charges of treason and execution only by his own natural death in 1530.

Henry immediately embarked on a substantial transformation of the residence and grounds, appropriate to his royal status. The king's first gardens at Hampton were inspired by the Renaissance developments in Italy. They were designed as a series of organized garden spaces beginning on the south side of the palace and extending down to the riverside. They included the privy, or private, garden, the pond garden or yard, and the mount garden.

Hampton Court Time Line

1525	Cardinal Wolsey offers Hampton Court to King Henry VIII as a gesture of kindness.
1638	King Charles I creates the Longford River to provide a new supply of water to the palace and gardens.
1660–1662	King Charles II introduces the Long Water Canal and the Great Avenues of Lime Trees.
1689–1694	King William has the privy garden redesigned. He creates the fountain garden on the east front of the palace, the wilderness and maze on the north end, and the banqueting house beside the Thames.
1694	Queen Mary dies. Work on the garden stops.
1697–1702	Work on the garden resumes.
1702	King William dies.
1764	Lancelot "Capability" Brown becomes the Royal Gardener at Hampton Court.
1768	Brown plants the Great Grape Vine.
1980s	Herbaceous border is planted along the Broad Walk.
1991–1995	Major privy garden restoration project is commissioned in 1991 and successfully completed in 1995.

FIGURE 12–1 The original palace at Hampton Court.

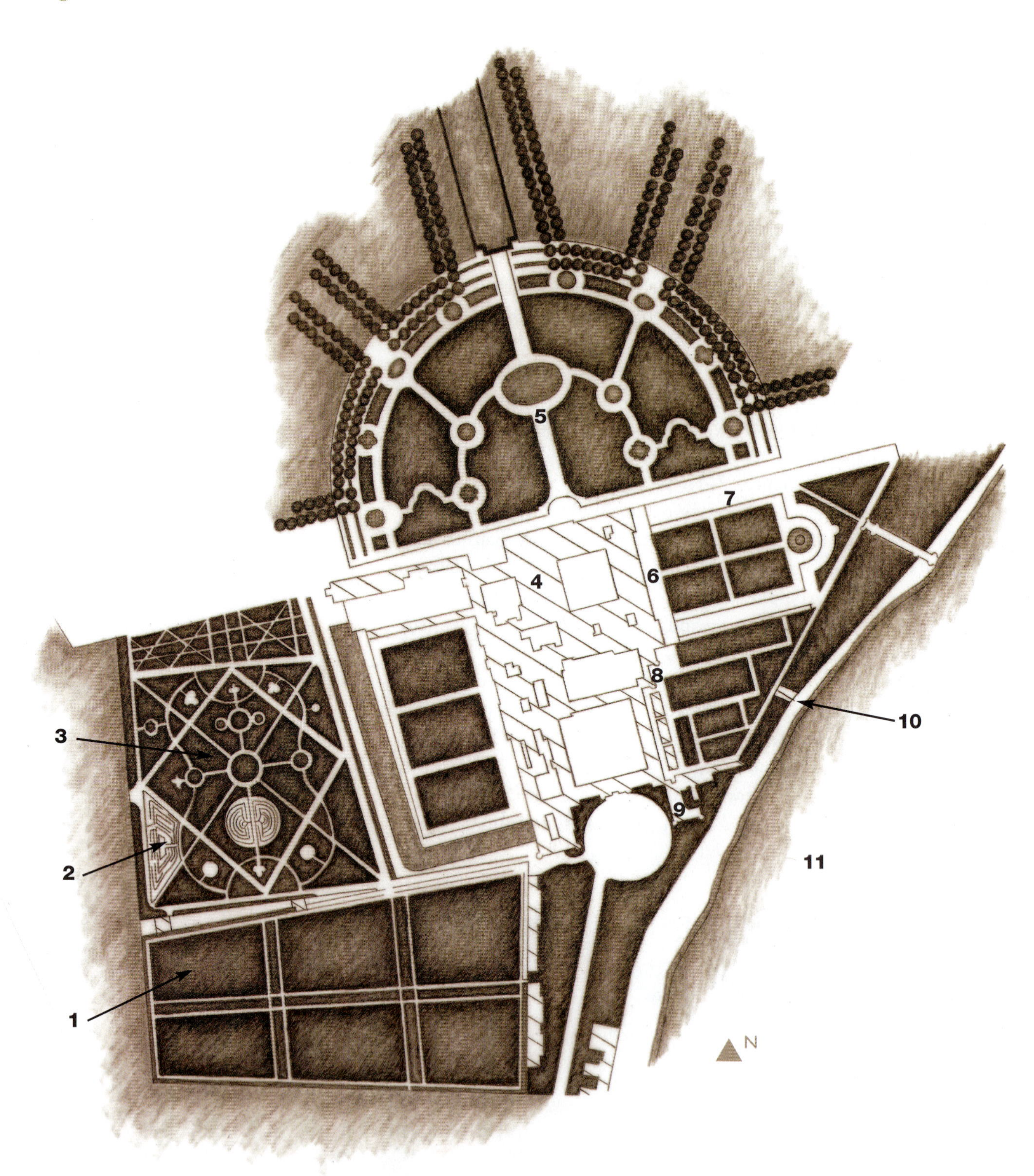

Hampton Court

1. The Tiltyard (Pater Kitchen Gardens)
2. The Maze
3. The Wilderness
4. Hampton Court Palace
5. The Fountain Garden
6. The Privy Garden
7. The Broad Walk
8. The Pond Garden
9. The Great Vine
10. Banqueting House
11. Thames River

Figure 12–2 King Henry's old pond yard converted into gardens by William and Mary during the seventeenth century.

Figure 12–3 View into the old pond yard from its entrance.

The privy garden to the south was reserved for the monarch and his chosen courtiers. It was divided into several enclosures by wooden posts and rails bordered by flowers such as roses, mint, sweet william, lavender, and thyme. The pond yard contained three walled ponds that held fish for the palace's kitchen. One pond was intended for breeding the fish, primarily carp, and the others likely held stock (Figures 12–2, 12–3). The walls enclosing the yard were decorated with painted stone beasts bearing shields displaying the king's coat of arms. Another of the king's creations, the mount, was built to provide views over the garden wall to the Thames and to the adjacent countryside. The mount was ascended by a winding path that spiraled upward to a gazebo, or summer house, at the top, offering splendid views of the river. A building called the "water gallery" was built on the riverbank as a place designed to receive visitors arriving to Hampton by boat. Most visitors traveled to Hampton by the river during Henry VIII's day, thus a pleasing reception building was an appropriate addition.

The area to the north of the palace was planted as an orchard and referred to as the privy orchard. To the northwest was a tiltyard, an area created to stage sporting activities and tournaments. The king was particularly fond of jousting and archery, which he was able to watch from any of the five towers created in the yard for spectators.

Henry VIII's daughter, Elizabeth I (1533–1603), developed a fondness for Hampton Court that continued to grow during her reign as queen (1558–1603). She found pleasure walking in the gardens and enjoyed viewing them from the windows of the palace. Elizabeth had a particular interest in the knot garden. She had one created near the south front of the palace. It consisted of dwarf boxwood laid out in an intricate pattern combined with herbs and flowers. The combination resulted in a splendid tapestry to be viewed and enjoyed from the palace windows above.

Few changes occurred in the garden until the reign of Charles I (1600–1649). In 1638, under Charles, a tributary of the Colne (a headstream of the Thames River) was diverted through Bushy Park to create the Longford River, providing a new supply of water to the palace and its gardens. Charles would later become a prisoner at Hampton, being confined there after the (second) civil war in 1648 until his execution in 1649, which abolished the monarchy and the House of Lords and declared England a commonwealth.

When Charles II (1630–1685) was restored to the throne in 1660 after his French exile, he displayed his affection for things French through the contributions he made to the gardens at Hampton, particularly in the eastern development. Charles had the great Long Water Canal dug as an act to impress his new bride, Catherine of Braganza (1638–1705). The canal is nearly a mile long and is 150 feet wide. Charles also created the Great Avenues of Lime Trees (lindens) planted on either side of the canal and along two avenues radiating eastward from a semicircle lined with the same trees in front of the palace (Figure 12–4). This work is clearly in the French spirit. The men who carried out the work, John Rose and André Mollet, each had experience with French design. The royal gardener, John Rose, had studied in France, and André Mollet was French born. Mollet was the former master of the great French landscape designer, André Le Nôtre. Le Nôtre was responsible for many of the seventeenth-century royal gardens belonging to Louis XIV. This explains the similarity between the

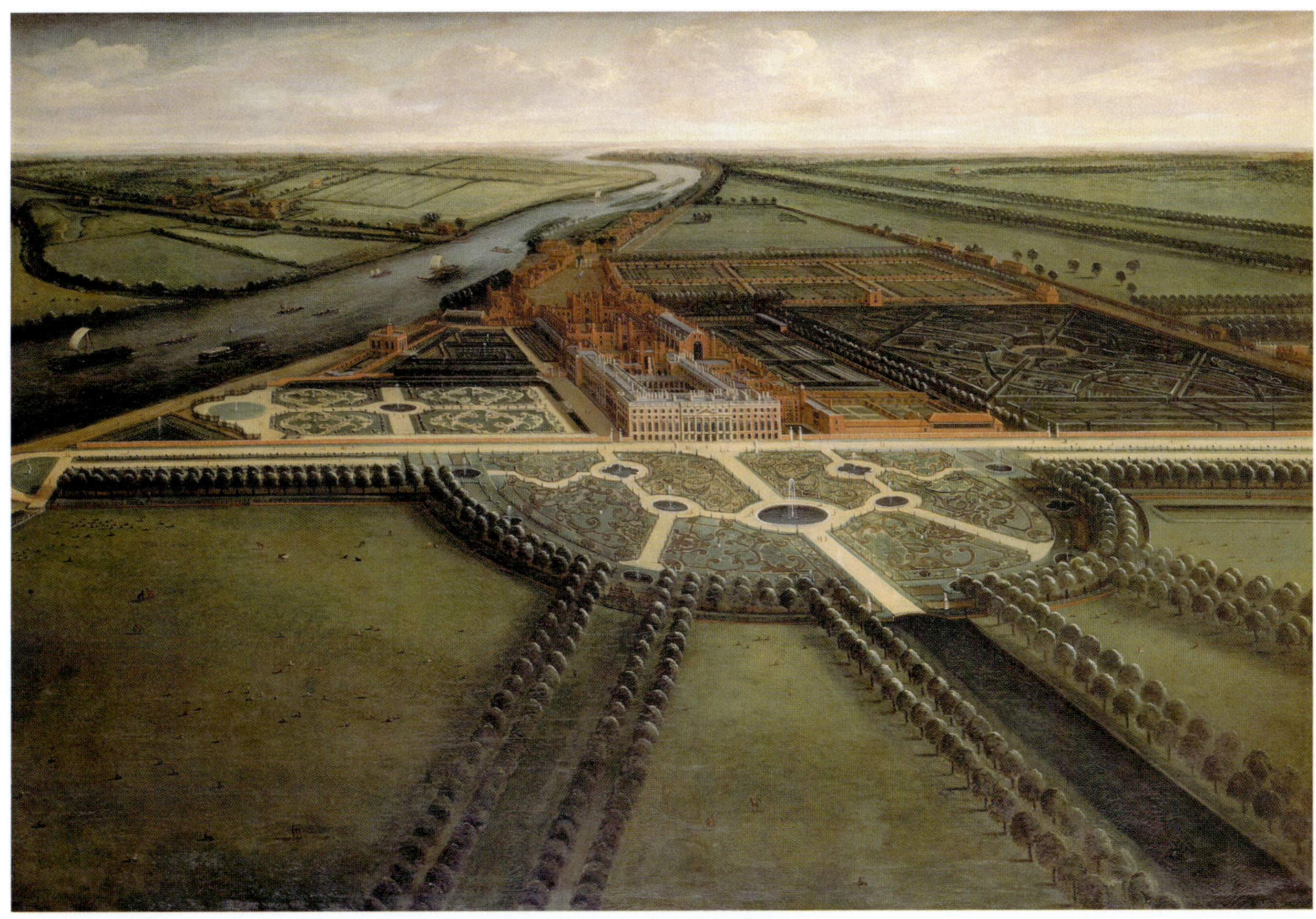

FIGURE 12–4 Hampton Court in the reign of George I, a detail from a painting by Leonard Knuff. Note the Great Fountain Garden commissioned by Charles II in the foreground of the image. The Royal Collection © 2004, Her Majesty Queen Elizabeth II.

layout of the East Gardens at Hampton and elements of Le Nôtre's work in France.

During the reign of James II (1633–1701), Hampton Court was neglected. When James was overthrown in 1689, the throne was offered to Mary (1662–1694), James's daughter, and Prince William (1680–1702), both of Holland. When William and Mary arrived at Hampton Court, they began to make changes to the gardens, beginning with the privy garden and then the east garden. The grassed areas of the privy garden were altered to include elegant *parterres* in the popular Gazon Coupé style. This technique, which originated in France, was accomplished by cutting turf into ornate patterns and filling the surrounding area with colored gravel or sand. Low hedge borders surrounded these turf patterns, and topiary yews and hollies were planted as accents within these borders.

At the southern end of the garden, toward the river's edge, decorative wrought iron panels were installed. These panels, called "Tijou screens,"

FIGURE 12–5 The privy garden as it appears today after a major renovation completed in 1995.

were designed by a French wrought iron specialist, Jean Tijou, who was responsible for all of the ironwork created for the palace. The design of the privy garden under William may have been the work of his gardener, George London, who had traveled to France and gained familiarity with the French style of garden design. The plan was actually drawn up by London's junior partner, Henry Wise, who also served the king as gardener. This garden in future years would become badly overgrown as a result of the eighteenth-century fashion for more natural-looking plantings favored by Queen Anne. The seventeenth-century design for the garden was reinstated through a major renovation project commissioned in 1991, (Figure 12–5).

On the east front of the palace an elaborate *parterre* garden was laid out with beds of boxwood designed in lacy patterns that surrounded 13 fountains. This garden, known as the fountain garden, was most likely created by Frenchman Daniel Moret, William's architect in Holland. It was designed to complement the new addition being constructed on the east front of the palace by Sir Christopher Wren. Splendid views of this garden could be enjoyed from the new state apartments. As planned, it would have been a great spectacle, though the fountains could never be made to work properly.

William's gardener, Henry Wise, and an engineer, Robert Alderly, both attempted to improve the fountains, but neither was successful.

Other changes to the gardens at Hampton Court by William and Mary included the alteration of Henry VIII's tiltyard. This area was divided into several kitchen garden plots that supplied the palace with fresh fruits and vegetables. In 1699, Henry's orchard to the north of the palace was converted into an area known as the "wilderness." The wilderness is not what its name might imply. It was designed as a large formal garden subdivided into smaller garden rooms by a series of hedge-lined pathways. The wilderness, likely designed by George London and Henry Wise, included a hedge maze in one of the "rooms" and a turf maze in another. Queen Mary was an avid plant collector and had several glasshouses constructed for her extensive collection of exotic plantings. These plants were moved outdoors for the summer months in the area that was formerly King Henry VIII's fishponds. Mary had each of the ponds drained and converted into gardens that accommodated her collection. Those gardens were later redesigned, but the Tudor walls remained, creating the series of walled flower gardens we see today. William continued to make changes to the gardens after Mary's death in 1694. In the privy garden, King Henry VIII's mount was removed. The water gallery was demolished and replaced with a small but elegant banqueting house situated beside the Thames.

In 1702, Mary's sister, Anne (1665–1714), became the queen of England. She made several modifications to the gardens, some pursued simply to distract her from the memory of her brother-in-law, William, whom she greatly despised. Other changes were made to reduce maintenance expenses. Most of the plantings at the fountain garden were removed and replaced with turf, partly because of Anne's dislike for boxwood, but also because the garden had been very maintenance intensive. All but five fountains in this garden were removed, and four more were later taken away under George I.

George I (r. 1714–1727) and George II (r. 1727–1760), the last of the reigning monarchs to live at Hampton Court, made few changes to the gardens. George III (1738–1820) was said to have asked Lancelot "Capability" Brown (1715–1783), popular designer of the eighteenth-century landscape park, to offer his expertise in making improvements to the grounds. Brown respectfully declined the commission. He later became the royal gardener at Hampton Court in 1764, but he made few changes or additions, aside from the planting of the Great Grape Vine in 1768, which thrives today. Changes in the nineteenth century included planting a herbaceous border along the Broad Walk at the east front of the palace. The herbaceous border as a style gained popularity in the late 1800s through the influence of garden designer Gertrude Jekyll and garden writer William Robinson.

Queen Victoria (r. 1837–1901) opened Hampton Court to the public in 1838. Today the gardens comprise 60 acres, with 750 acres of parkland. They are impeccably maintained by a full-time staff of 41 gardeners, far fewer people than were under the supervision of Henry Wise in the 1700s, when several hundred men cared for the grounds. Now of course technology has increased the efficiency of garden maintenance. Nevertheless, the scale of work today is still impressive. Over 140,000 bedding plants are raised on site each year to supply the gardens, and more than 200,000 spring bulbs are planted annually.

The impressive maintenance of the gardens is regarded as a team operation. The ongoing challenge is a balance between maintaining historical integrity and creating a beautiful garden that appeals to the modern visitor. The staff at Hampton Court skillfully achieves that balance.

Chapter 13

Landscape Expression: The Early-Eighteenth-Century English Landscape

FIGURE 13–2 View from the garden out to the Oxfordshire countryside.

carved from the woods into which statues of classical gods are featured), a bowling green, and three geometric basins that, under Kent, would be a part of the Vale of Venus.

Bridgeman's plan was the first attempt at Rousham at England's new practice of gardening. Bridgeman tried to create a dialogue between the garden and the countryside, encouraging peaceful views to the adjacent river and surrounding landscape. However, Bridgeman's layout was still very geometric and somewhat rigid, with sharp angles directing garden views. This approach likely compromised the intended serenity of the visual experience.

Under Kent, the garden became two distinct but connected experiences, one classic and very private or personal garden and another rather picturesque garden with views directed to the countryside (Figure 13–2). Kent retained the basic plan of the grounds laid out by Bridgeman but reshaped and improved the garden to highlight nature. He developed garden forms that allowed the eye to move more freely across the landscape. He also was effective at creating a particular mood in the garden by directing and framing views of the countryside that set the tone for the peaceful image of the surrounding landscape that he intended.

FIGURE 13–3 "Eye-catchers" were placed out in the fields to attract views.

Kent's transformation of the garden involved several architectural interventions designed as framing devices that were meant to capture views of different areas of the landscape. At the same time, these constructions present themselves as part of the scenery viewed from other areas in the garden (Figure13–3). The countryside was drawn into the designed space and became a vital element in the total experience of the garden—in essence, it was part of the garden. Kent was also especially successful at organizing perspectives in ways that visually extended the site, creating the perception of an infinite landscape on what was actually a rather small property. His appreciation for the experience of a **borrowed view** (scenic views beyond those designed within the garden's boundaries, and intended to contribute to the overall design) was acquired during his tenure in Italy. The villa gardens of Renaissance Italy were typically sited in a way that offered views of the surrounding countryside. It was an important part of the villa experience. At Rousham, Kent succeeded in localizing this Italian experience by capturing views of the Oxfordshire countryside and making that visual experience an important element in the garden. Kent's garden was filled with symbolic sculptures that conveyed acts of heroism and celebrated political liberties.

FIGURE 13–4 A view south up the bowling green toward the house.

These garden elements created a sort of historic review of General Dormer's lifelong career defending British liberty. Hence, Kent's architectural interventions are not simply classic decorations but, rather, classic emblems representing many associative ideas relevant to Dormer's life.

The garden experience begins at the rear of the house with a long view down across a **bowling green** (a rectilinear area of lawn that helps define a visual axis, similar to the French *tapis vert,* Figure 13–4). At its end is a copy of an antique sculpture depicting a lion engaged in fierce combat with a horse (Figure 13–5). This, one of the many sculptures in Dormer's personal collection, seems to suggest a symbol of great power, bravery, and strength, all attributes of a good soldier. The sculpture, perched on the very edge of a bank that descends to the river Cherwell, has an open view of the Oxfordshire countryside beyond. The garden path from the bowling green leads to the site of Kent's Praeneste Terrace, named after the tiered ruins of the ancient arcade located on a hillside at Palestrina in Italy. It is from this elegant, ancient structure that Kent gained his inspiration for the single-arcaded feature at Rousham. As Patrick Eyres points out in his essay (1985), both the ancient Roman ruin (the Temple of Fortune at Praeneste) and Kent's abbreviated version at Rousham were a vantage point for extended views—in Rome, across the Campagna, and at Rousham, across the Oxfordshire countryside. The association of the nearby Dying Gladiator sculpture (located on a terrace directly above the praeneste) suggests an intended comparison between General Dormer and the great Roman soldier and emperor Marcus Aurelius (Figure 13–6). Aurelius was said to have taken retreat at Praeneste prior to his death, which occurred during his military service.

Classical associations provide a deeper meaning to the garden at Rousham. The Lion and the Horse and the Dying Gladiator sculptures both seem to contribute to a theme of battle and death, while Kent's Vale of Venus seems to counter that theme as a symbol of love and life. Kent redesigned the Vale, a sequence of descending pools of water originally laid out by

FIGURE 13–5 The antique sculpture of a Lion Devouring a Horse at the north end of the bowling green.

FIGURE 13–6 The Dying Gladiator sculpture by Peter Sheemakers.

FIGURE 13–7 The Venus Vale, originally laid out by Charles Bridgeman and redesigned by William Kent.

Bridgeman, by incorporating two rustic cascades along the sloping lawn (Figure 13–7). Venus, goddess of love, is the presiding figure here, emphasizing life and fertility. A sculpture of the mythological goddess stands atop the upper cascade.

After Kent, a movement toward a greater "naturalness" in gardens began. Landscapes would become even more simplified. Nature would be idealized and celebrated for its beauty. Classical associations in gardens would be few. Kent's accomplishments at Rousham anticipated that transition. His work successfully united nature and designed garden as one seamless composition.

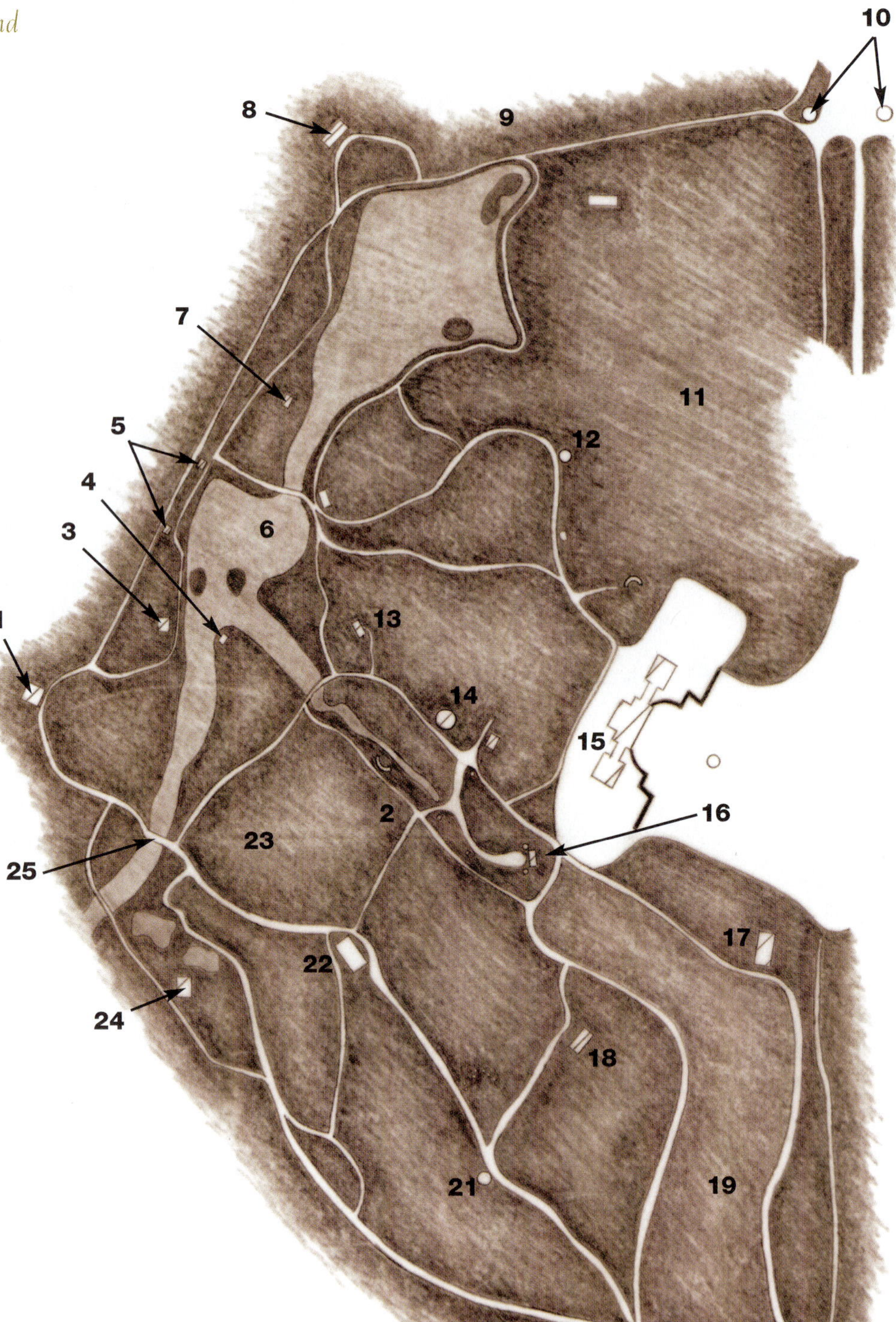

Stowe

1. Temple of Friendship
2. British Worthies
3. Pebble Alcove
4. Congreve's Monument
5. Lake Pavilions
6. Octagon Lake
7. Hermitage
8. Temple of Venus
9. Ha-Ha
10. Boycott Pavilions
11. Western Garden
12. Rotundo
13. Doric Arch
14. Temple of Ancient Virtue
15. Stowe House
16. Grotto
17. Temple of Concord and Victory
18. Queen's Temple
19. Grecian Valley
20. Fane of Pastoral Poetry
21. Cobham Monument
22. Gothic Temple
23. Eastern Garden
24. Chinese House
25. Palladian Bridge

Stowe

BUCKINGHAMSHIRE, ENGLAND

Of all the gardens in England, Stowe in Buckinghamshire is the best known and most frequently visited. Writers, poets, and garden historians have described it throughout its 300-year history as an extraordinary work of art and nature combined. Yet Stowe is more than a beautiful garden. It was designed as a statement of the political and moral principles of its maker, skillfully conveyed in physical form.

The garden as it exists today was begun by Sir Richard Temple (1675–1749), the 1st Viscount of Cobham. Its program developed as an expression of political opinion conveyed in an itinerary of garden buildings, temples, and sculptural elements. Viscount Cobham was a Whig politician. The Whig Party was an opposition group that was rewarded political freedom as a result of the Glorious Revolution (1688). The group, which ultimately formed the Liberal Party in England, supported the supremacy of parliament over the king. The Whigs achieved control of the government of England in 1714 upon the accession of King George I. Cobham, the party's effective leader, was a highly regarded military officer who was appointed lieutenant general of the British army in 1710. He was raised to the rank of a noble, becoming Viscount Cobham in 1717.

Lord Cobham inherited the Stowe estate in 1697. As the Whig Party grew, so too did Stowe, which became the center of political activity for the

Stowe Time Line

1678	Sir Richard Temple, 3rd Baronet, begins making improvements on the property, creating a vineyard, a kitchen garden, and an orchard.
1680	Temple creates formal pleasure gardens.
1711	Temple's son, Richard, the 1st Viscount of Cobham, begins making improvements on the Stowe estate.
	Charles Bridgeman is hired as Cobham's garden designer.
	Sir John Vanbrugh is hired as architectural advisor.
1713–1730	Formal gardens are laid out by Charles Bridgeman and Sir John Vanbrugh for Cobham.
Late 1720s	William Kent first arrives at Stowe as Cobham's architectural consultant.
1731	William Kent returns to Stowe and gains garden commissions.
1730s–1740s	East gardens are laid out by Cobham.
1739	James Gibbs is appointed Cobham's chief architect.
1741	Lancelot "Capability" Brown is appointed head gardener and begins making contributions to the garden.
1749	Richard Grenville Temple, later Earl Temple, takes possession of Stowe.
1750	Lancelot "Capability" Brown leaves Stowe. Temple hires Richard Woodward as head gardener, and begins making alterations to the existing landscape.
1848	The contents of the house are sold to pay creditors.
1922	The Stowe house and garden become the property of the Stowe School.
1990	The gardens become the property of the National Trust.

Whigs, the core group of which included many of Cobham's family and friends. It was essentially the place from which England was governed until 1760. The first gardens to be created at Stowe were not by Viscount Cobham, however, but by his father, Sir Richard Temple, 3rd Baronet. Sir Richard began making improvements on the property in 1678, first by constructing a home, still part of the estate, and then by planting gardens and orchards. In 1680, he created a small, terraced pleasure garden, to complement his newly built home. The garden had a formal character, consistent with the design style of most gardens created in England at that time.

Cobham did little to change the gardens until his financial position grew to a level that permitted him to do so. By 1711, he was able to begin making improvements, expanding his father's early work and establishing an estate commensurate to his position in society. He assembled a team of accomplished designers to help him realize his vision. Charles Bridgeman, who held the position of royal gardener, was hired as Cobham's garden designer, and Sir John Vanbrugh, a friend and member of the prestigious Kit-Kat Club, to which both Bridgeman and Cobham belonged, was hired as his architectural advisor.

The gardens at Stowe, today covering approximately 400 acres, can be viewed as three distinct areas. The first, the main garden vista, occurs along a rolling lawn that follows the southern axis, the location of the original seventeenth-century garden terraces. The second area, to the west of this principal axis, was gardens laid out by Charles Bridgeman for Viscount Cobham between 1713 and 1730. These gardens today do not reflect the original layout. They were redesigned to reflect the naturalistic style in gardening that dominated England during the mid-eighteenth century. The third area is the east gardens, commissioned by Cobham in the 1730s and 1740s. These gardens generally appear as they did in the 1740s. They evolved as three distinct areas—Elysian Fields, Hawkwell Field, and the Grecian Valley—each with its own unique characteristics.

Elysian Fields exists within a narrow valley that parallels the main north-south vista from the house. A stream winds through the valley, its banks serving as a canvas against which numerous classical structures by William Kent have been artistically positioned. Kent, first arrived at Stowe in the late 1720s as an architectural consultant hired to advise on improvements to the house. His role was extended in 1731 to include full-time commissions in the garden. The significant buildings and temples Kent designed express the political and philosophic ideas of Lord Cobham and his circle of friends or fellow Whigs. Hawkwell Field was conceived of as a working farm, with grazing animals roaming across the hayfields. Lord Cobham valued it for its peaceful appearance, interrupted only by symbolic buildings that contribute to the garden's political agenda. The Grecian Valley, the last area to be developed, lies at the northeast corner of the garden. Its pastoral composition represents nature perfected in a style that became popular under the guidance of Lancelot "Capability" Brown.

For today's visitor, it often is difficult to perceive Stowe as a garden, but in eighteenth-century England, the idea of a garden was more than a horticultural composition. The early eighteenth-century style was influenced by the upper-class society's desire for gardens that were intellectually satisfying. One distinguishing characteristic of the upper classes was a classical education, and the ideas presented in the garden typically called upon that knowledge by making symbolic references to the classical world.

The gardens at Stowe were created along three distinctively different programs of development. Their evolution reflects changing philosophies in English garden theory and the "idea" of the garden. The first phase was relatively formal, clearly faithful to classic precedents. The second phase, still somewhat formal, transformed the estate into a grand statement of the owner's status. In the final stage, the new, eighteenth-century vogue for a more natural-appearing garden was created. Nature was rediscovered and celebrated for its beauty. New developments in the garden were less imposing and more in harmony with nature.

The first phase of development began in 1714 and continued for six years. The work, guided by the expertise of both Charles Bridgeman and Sir John Vanbrugh, advanced with a formal character. Axes were established as the main lines of the garden, and formal avenues of trees were planted to delineate pathways and boundaries. A barrier ditch devised by Bridgeman, the ha-ha, was developed to enclose the garden without the need for a fence. It prevented livestock from either entering or leaving the property's boundaries and also allowed for uninterrupted vistas of the adjacent countryside (Figure 13–8).

FIGURE 13–8 Bridgeman's ha-ha was created along the southern boundary of the property.

Shortly after his promotion to a viscount, Cobham began to consider expanding his garden to create something even more magnificent and commensurate to his new status. Between 1720 and 1724, the gardens were enlarged, mostly to the west, based on a master plan created by Bridgeman. Bridgeman's greatest challenge was to create a coherent layout of separate garden areas within an oddly irregular site boundary. He did this by developing a system of both visual and physical links that included beautiful garden vistas and tree-lined walkways that moved the visitor around the site.

Architectural elements were constructed to capture one's attention and to direct or frame views. Two buildings, named the "Lake Pavilions," designed by Sir John Vanbrugh, were constructed to frame the avenue to the south. A **rotundo** (a round, domed building, often surrounded by a colonnade) also designed by Vanbrugh, was built upon a mound at the intersection of two main avenues forming the western boundary. Though organized and connected by straight lines, the expanded garden did not really appear rigid or formal. Views directed to open vistas and the natural countryside became part of the garden experience, connecting it with nature and making it seem less formal.

The Temple of Venus and the Hermitage were the last two buildings to be added to the western garden at Stowe. These buildings were William Kent's first commissions in the garden upon his return in 1731. His first garden building, the Temple of Venus, was designed as a miniature Palladian villa that overlooked the 11-acre lake. Inside, the central room of the temple was decorated with risqué murals painted by Francesco Sleter, a Venetian artist. The murals depict scenes from Spencer's *Faerie Queen,* a story about a man and his adulterous wife. These murals no longer exist, having been painted over during one of the restorations to the temple. The Hermitage, near the Temple of Venus, continues Spencer's story. The betrayed and disgusted husband in the story retreats to a cave, here represented by Kent's building.

Having completed the west garden, Cobham set out to develop his garden further by expanding to the east, balancing the composition on either side of the central vista. The east garden developed as a scenic landscape with political undertones that made reference to Cobham's libertarian values. It also established a design precedent for the new vogue of English gardening, with a more natural, parklike layout.

The plan of the new east garden is credited to Kent. The first area he designed was a narrow valley lying parallel to the central vista on axis with the house. This area became known as Elysian Fields, a splendid garden so named for Elysium, the mythological paradise where the virtuous reside after death. Within this garden, Cobham created a setting for a carefully devised program of ideas, a symbolic garden that conveyed the political philosophies and allegiances held by him and his Whig Party. On the west side of the valley a significant building, the Temple of Ancient Virtue, was constructed in 1736. Designed by Kent, it contains statues representing ancient Greek heroes who were known to possess the qualities of leadership that Cobham and his fellow Whig patriots desired for England's government. To the south of the temple stood the Temple of Modern Virtue, built as a ruin. Among its rubble stood a single headless figure, which may have represented the English prime minister, Sir Robert Walpole (Clarke et al. 1997). Though there is no evidence to substantiate this assumption, its former location seems to present a plausible comparison between the degenerate state of government during Cobham's time and the noble accomplishments of the ancient Greek heroes.

On the east bank of the river lies another temple designed by Kent, the Temple of British Worthies, built in 1734 as a semicircular construction containing 16 stone busts in niches (Figures 13–9, 13–10). Half of the figures represent those revered by Cobham for their philosophies, and the other half were those honored as persons of action. The thinkers or intellectuals represented in Cobham's scheme included writer William Shakespeare, scientist Isaac Newton, and philosopher Francis Bacon. These were popular British figures and obvious selections for a hall of "worthies." The respected "figures of action," as he called them, were less obvious. Each of the individuals represented was selected for an action that in some way contributed to Cobham's political mission. For example, a member of Parliament, John Hampden, was honored for supporting Cobham by voting against Walpole's proposed excise tax.

At the bottom of Elysian Fields, situated on an island in the Octagon Lake, is a monument dedicated to dramatist William Congreve (1690–1729), an old friend of Lord Cobham's. The memorial was designed by William Kent in 1736 as a stone monkey sitting atop a pyramid. The monkey appears to be staring at himself in the mirror, alluding to the very nature of Congreve comedies, which were often created as mirrors of society.

FIGURE 13–9 The Temple of the British Worthies, designed by William Kent in 1734.

The land to the east of Elysian Fields was enclosed by Bridgeman's ha-ha in the early 1730s. By the end of the decade Cobham set out to expand the garden into this pasture area, Hawkwell Field, a move that united the garden and the countryside beyond. This area continued Cobham's theme of political expression, but the setting was more pastoral. Animals grazed among buildings, effectively contributing to the rustic scenery.

FIGURE 13–10 View of the Temple of the British Worthies, John Claude Nattes, 1805. Buckinghamshire County Museum and Stowe School, England.

Figure 13–11 The Gothic Temple and Queen's Temple, John Claude Nattes, 1805. Buckinghamshire County Museum and Stowe School, England.

The architect guiding this area's development was James Gibbs, who was first introduced to Cobham by Charles Bridgeman shortly after Sir John Vanbrugh's death in 1726. Gibbs returned to Stowe full time as Cobham's architect in 1739, after the extensive period of development in the 1730s credited to Kent. Gibbs was responsible for the design of a number of major buildings in the garden, primarily in the eastern development. He did, however, design two baroque-style pavilions, called the "Boycott Pavilions," (c. 1728), at the western entrance to Stowe to announce the new grand entry to the park.

Gibbs' first project upon his return to Stowe was the Temple of Friendship, commissioned by Lord Cobham as a retreat or meeting place to gather with his closest friends and allies. It was named to commemorate the visit of Frederick, Prince of Wales, in 1737. Frederick was greatly respected by the Whig patriots for sharing many of their political philosophies. The temple is situated on the east end of the southern portion of Hawkwell Field, creating an effective balance with the Temple of Venus on the west side. While both the Temple of Venus and the Stowe estate could be seen from the Temple of Friendship, a more significant connection was made with the Lady's Temple (later the Queen's Temple). Designed by Gibbs and constructed between 1744 and 1748, this garden building situated at the opposite (north) end of Hawkwell Field directly faces the Temple of Friendship. Its original name, the Lady's Temple, is derived from its intended use as a place for Cobham's wife to spend private time with her closest friends.

The Gothic Temple, or Temple of Liberty, perhaps the most impressive building designed by Gibbs, represents a significant addition to Cobham's politically inspired garden itinerary (Figure 13–11). The Gothic style was chosen for being representative of ancient English liberties. As Robinson points out, to the Whigs, "Saxon" or "Gothic" was associated with the freedom of an

FIGURE 13–12 The Palladian Bridge, by James Gibbs, at the bottom of Hawkwell Field.

Englishman (Robinson 1994). Parliamentary representation and trial by jury, for example, were both believed to be Saxon or Gothic in origin. It was these very rights and privileges that were protected in both the civil war and Glorious Revolution, and the very principles that Lord Cobham and the Whig patriots were dedicated to preserving. The temple stands proudly at the top of Hawkwell Field, surrounded by a grove of trees. It is the climax of the political theme developed in the garden.

At the bottom of Hawkwell Field is the Palladian Bridge (Figure 13–12). This bridge was the second of three such bridges to be built in England during the eighteenth century. The first was constructed at Wilton in 1737 for the 9th Earl of Pembroke, who along with architect Robert Morris developed the design. The third was created near the city of Bath at Prior Park, with Thomas Pitt being credited for its design. The Stowe Bridge was created by Gibbs in 1738. While similar to the Wilton Bridge, there are several distinct differences. The Stowe Bridge serviced a carriage road, thus it was built with ramps on either end, unlike the other two bridges, which were constructed with steps. The bridge also is set lower to the water and thus appears less dramatic, yet it is richer in detail than the bridges at Wilton and Prior Park, which compensates for its lack of elevation.

Along the south side of the Octagon Lake is a garden seat situated within a shelter called the "Pebble Alcove." Within the alcove is a pebble mosaic, representing Lord Cobham's coat of arms. Beneath Cobham's emblem are the words *Templa Quam Dilecta* ("How beautiful are thy temples"), words that aptly describe Stowe. The shelter, believed to be designed by William Kent sometime before 1739, has been twice restored, in 1877 and 1967.

The last section at Stowe to be developed by Cobham was the Grecian Valley. This area, originally an open pasture, extends to the northwestern portion of the property. Lancelot "Capability" Brown, who began work in this area of the garden in 1747, transformed it into a splendid green vale. The transition of this treeless pasture into a wonderfully picturesque valley involved a substantial amount of work. Over 23,000 cubic yards of soil were moved. Mature trees were transplanted from other areas of the site to create groves of evergreens and hardwoods on either side of the valley. This "ideal landscape," both scenic and seemingly natural, would serve as an example of Brown's developing style of landscape design. The valley is named for the Grecian Temple, now called the "Temple of Concord and Victory," begun in 1747. It is the largest of Stowe's temples, believed to have been originally designed by Kent, though Cobham's nephew and heir, Richard Grenville, later Earl Temple, may have had some influence on the final design. The temple, though designed with inaccuracies, was intended to represent the building style of ancient Greece. In the second half of the eighteenth century (1763), Earl Temple, celebrating the victory of the Seven Year's War, renamed the temple "Concord and Victory." Inside, a series of 16 terra-cotta medallions celebrates the British victories that gained them control of North America.

The tallest of Stowe's monuments is the Cobham Monument located between Hawkwell Field and the Grecian Valley. Completed just prior to Cobham's death in 1749, the 104-foot tower was dedicated to Lord Cobham by his wife. Originally designed by architect James Gibbs, the plans were later modified by Lancelot "Capability" Brown, who directed its construction in 1747. From its uppermost viewing platform, one has a commanding view of the garden. A statue of Cobham stood proudly atop the column until 1757, when it was struck and destroyed by lightning.

Earl Temple, the nephew of Lord Cobham, took possession of Stowe in 1749, shortly after his uncle's death. Temple had enjoyed a fine education, spending much time abroad, where he acquired a taste for art and architecture. Anticipating his inheritance, he had sufficient time to develop his own ideas for making improvements at Stowe. Following the departure of "Capability" Brown in 1751 as head gardener, Temple began to make alterations to the existing landscape. A new gardener, Richard Woodward, was employed, but it is believed that Temple himself was responsible for most of the designed changes. His main focus was to naturalize any of the remaining formal vocabulary in the garden, which included softening the outline of Octagon Lake. Other major changes included moving the Lake Pavilions farther apart to open up the south vista (1764) and building the Corinthian Arch, designed by Temple's cousin, Thomas Pitt, in 1765 as a focal point on the distant horizon.

The family fortune declined at the hands of Temple's successors. In 1848, most of the contents of the estate were sold to pay creditors. Despite these financial difficulties, Stowe remained in the family until 1921, when it was sold to a private investor who held it for a short time. The house and garden became the property of a newly developed school, fittingly called the "Stowe School," in October 1922. The beautiful historic buildings and landscape are central to the educational ideals of the school as expressed by the school's founders in 1923. J. F. Roxburgh, upon his appointment as Stowe's first headmaster, stated, "If we do not fail in our purpose, every boy who goes out from Stowe will know beauty when he sees it, all the rest of his life" (Robinson 1994).

Stourhead

1. Stourhead House
2. Stable Yard
3. St. Peter's Church
4. Bristol High Cross
5. Palladian Bridge
6. Temple of Apollo
7. Pantheon
8. Gothic Cottage
9. Grotto
10. Temple of Flora

Stourhead

WILTSHIRE, ENGLAND

Stourhead in Wiltshire, three miles northwest of Mere, represents the idyllic English eighteenth-century garden. Its creator, Henry Hoare II, a wealthy London banker, built this fine English landscape and also helped many wealthy landowners create their own. During the eighteenth century, many ambitious estate owners sought to align themselves with the developing trend for creating picturesque landscapes to represent their status in society. The problem that many encountered was poor budgeting. An enormous amount of money was required for such projects, especially the type of landscapes that most wealthy landowners envisioned. To help finance their grand projects, many landowners turned to Henry Hoare (1677–1725), whose family had been in banking since 1672. This surge in development and the requirement for loans brought new wealth to the Hoare family, already a successful pedigree.

The Hoare family dynasty began with Richard Hoare (1648–1718), who pioneered the creation of a commercial banking system in London in the late seventeenth century. He served as London's representative in Parliament from 1709 to 1713 and was lord mayor of London in 1712. Under Queen Anne, he was knighted. Sir Richard Hoare also was a founder and one of the original directors of the South Sea Company, a government-sponsored trading business. The company was created to relieve the Crown's debts, which had risen considerably during Charles II's administration. The trading company gained in popularity as it drew profits from government monies collected on imports. Its stock share rose beyond the company's value, and many of the directors, fearing an eventual correction, sold their shares. Richard died in 1718, two years before this speculation began. His sons, Benjamin and Henry, were able to sell their shares in time to make a significant profit. Further selling precipitated, and the shares eventually plummeted, causing numerous bankruptcies, the bursting of the "South Sea bubble," as it was called. Having acquired a sizeable fortune, Richard's son Henry was thus able to create a spectacular estate of his own called, "Stourhead."

Henry hired architect Colin Cambell, leader of the Palladian revival in England, to design his home. Cambell was already associated with Hoare's family, having been appointed to the position of Deputy Surveyor of the Royal Works by William Benson, Hoare's brother-in-law, who held the position of surveyor general. Henry died in 1725, at which time the house was nearly complete. His widow, Jane Benson, continued to live there until her death in 1741. It was Henry Hoare II (1705–1785), nicknamed "the Magnificent," who completed the Palladian style villa begun by his father and developed the landscape around the grounds, considered then and now one

Stourhead Time Line

1744	Henry Hoare II "the Magnificent" begins to make improvements to the estate inherited from his father, Henry Hoare Sr. Henry Flitcroft is hired as Hoare's architect.
1785	Sir Richard Colt Hoare inherits Stourhead upon his grandfather Henry's death.
1785–1838	Colt Hoare introduces into the garden numerous exotic species from both Asia and the Americas.
1894	Sir Henry Hoare, 6th Baronet, inherits the estate.
1894–1936	Dead and declining trees and shrubs are replaced. A vast number of rhododendrons and azaleas are added to the garden.
1936	Henry, 6th Baronet, without an heir, bequeaths the house and gardens to the National Trust.
1946	The National Trust takes over the administration and upkeep of the landscape garden.
1953	High winds associated with a heavy storm destroy many large trees in the woodland.
1960s	The trust is awarded grant funding to make repairs in the garden.
1978	The trust establishes a Conservation Plan detailing criteria for long-term management of the garden.

of the finest in all of England. Henry II succeeded his father as head of their flourishing Fleet Street bank business. From a very young age, he enjoyed a luxurious and an unrestrained lifestyle and acquired a taste for fine art. He traveled to Europe, where he became familiar with Renaissance artists, especially the work of Claude Lorrain and Nicolas Poussin. His life as a youth was privileged, but his adult years were plagued by a sequence of unfortunate and terribly tragic events. He was twice married, first to Ann Masham, who died while giving birth, and then to Susana Colt. Colt died in 1743, leaving Henry with a 13-year-old son and two daughters, ages 11 and 6. His life of misfortune continued, as each of his children died before him.

Shortly after his wife's death in 1743, Hoare began to make substantial alterations at Stourhead. He hired Henry Flitcroft (1697–1769) as his architect. Flitcroft was responsible for the architectural interventions as well as for the initial planning of the lake, around which the landscape would be developed, but it was Henry, inspired by his love of art, who was responsible for most of the design of this scenic landscape park.

Flitcroft continued to assist Hoare at Stourhead for a period lasting over 20 years. In his first assessment of the site, he advised him to transform the stream and small pond in the valley into a grand lake to serve as the centerpiece in his design. Following his advice, Hoare constructed a dam, causing the valley to flood. This created the lake that Flitcroft had imagined. A path was laid around it, leading to a series of stations established as part of an itinerary of classical associations.

Numerous "staged scenes" along a predetermined path convey Hoare's interest in the antiquities. His design gives physical form to the seventeenth-century paintings he so admired and as historian Kenneth Woodbridge (2001) points out, life to the world described in his collection of classic texts such as Virgil's *Aenid* (30–19 B.C.). The first of a series of buildings created in the garden was the Temple of Ceres, built in 1744. This temple, designed by Flitcroft and now known as the "Temple of Flora," stands at the head of the lake and was intended as the first to be approached on the circuit. Above its entrance is a quotation from Virgil's text (VI, 258), *"Procul, o procul este profani"* ("Begone, you who are un-initiated or unworthy begone!").[1] These words seem to serve as an initiation to visitors, mentally preparing them for the garden's classical context. From here the path travels along the lake to the opposite side and then descends into a dark grotto. Inside, a sleeping nymph reclines over a cascade of spring water. An arched opening offers a view out across the lake, back toward the Temple of Flora. This construction, also by Flitcroft, further implies Virgilian context, making reference to another incident in the story of Aeneas' journey and the founding of Rome.

From here the visitor descends to another cavelike construction and finally travels back up to the main path along the lakeshore. The next garden view to be experienced along the path seems to have been inspired by

---

1. Virgil's "Aeneid" in *The Stourhead Landscape,* Kenneth Woodbridge, The National Trust, 2001 p. 19.

FIGURE 13–13 The Pantheon, by Henry Flitcroft, is modeled after the great Pantheon in Rome.

Lorrain's *Coast View of Delos with Aeneas* (1672, now in London's National Gallery of Art). In Lorrain's painting, Aeneas stands in the foreground with his father, his son, and a priest. All are looking toward a temple that appears much like the Pantheon in Rome. Rome is the great city that Aeneas is destined to establish in the context of Virgil's *Aeneid.* We know that Hoare was particularly fond of Claude Lorrain, owning at least one of his paintings, which makes the assumption even more likely that the painter's work inspired the garden especially this particular vista. The Pantheon building represented in the garden was designed by Flitcroft and is perhaps one of his greatest works (Figure 13–13). Inside, sculptures have been placed in niches in the surrounding walls. Olin suggests that the sculptures which represent ancient gods and goddesses were chosen by Hoare to represent aspects of his own life, particularly his experiences of love and his endless confrontations with death. Lead figures of Venus and Bacchus, for example, stand in niches on either side of the entry portico, announcing this theme. The goddess Venus represents love and Bacchus, god of the harvest, represents the cycles of nature and the mystery of rebirth, or life after death (Olin 2000, 269).

From the Pantheon is a splendid view across the lake to the village of Stourton, and the medieval church of St. Peter's. Foregrounding the church is an architectural sculpture called the "Bristol High Cross," dating from 1373.

FIGURE 13–14 A turf-covered bridge of stone construction with five Palladian arches spans the lake.

Once a city monument marking the independence of Bristol's merchants, it was later regarded as a traffic hazard and removed from the city's main street. Hoare salvaged it for use in his garden, placing it in a significant location where it encourages views toward St. Peter's Church and the village of Stourton. An elegant turf bridge defined by a series of five Palladian arches (Figure 13–14 and Figure13–15) adds to the picturesque quality of the view, again inspired by Lorrain's *Aeneas at Delos.* To the west, standing atop a hill, is the Temple of Apollo, designed by Flitcroft in 1765 (Figure 13–16). From the temple one can look down upon the landscape, visually retracing one's route through the garden. The view from this elevated prospect is extraordinary. As English writer, Joseph Spence, remarked, "From it you take in all the chief beauties of the place."[2]

FIGURE 13–15 View of the Garden at Stourhead, looking toward the Temple of Flora, the Bristol Cross, and the Palladian Bridge, c. 1775, Coplestone Warre Bampfydle.

What was it that inspired Henry Hoare to create such an extravagant garden? Perhaps he was inspired by the prestige of having such an elaborate place, one that people from all over the country would be eager to see. This makes perfect sense, since visiting gardens was a fashionable activity in the late eighteenth century, popular among those with "taste." It also could be

2. Joseph Spence, *Stourhead Landscape Garden,* The National Trust, 2000, p. 18.

FIGURE 13–16 The Temple of Apollo, by Henry Flitcroft, offers a panoramic view of the garden from its hilltop location.

that Stourhead was Henry Hoare's therapy, a diversion from the tragic events that had occurred in his life.

By the late eighteenth century, the essence of gardening throughout England had changed. The Napoleonic Wars (1700–1815) caused England considerable financial hardship and temporarily halted most activities related to cultural advancement. Since landscape gardening, like other arts, seemed more attractive in times of peace and prosperity, the psychological and economic factors of war initiated a change in both the extent of and the idea for garden development. Wealthy landowners became more private and less extravagant in the development of their properties. Their interests seemed to focus more on the planted landscape. English horticulture was making substantial advances. Plant collecting had become quite popular with aristocratic amateurs by the end of the eighteenth century. Gardens became grounds for both scientific study and display.

At Stourhead, we see evidence of this changing attitude toward gardening. Henry Hoare's grandson, Sir Richard Colt Hoare (1758–1838), inherited the house and garden upon his grandfather's death in 1785. His inherited wealth enabled him to continue to build and refine the estate. He, like his grandfather, approached the garden from the perspective of an artist, but Colt Hoare desired a more simplified and natural-looking landscape. He aimed to naturalize the scenery and broaden its aesthetic appeal through the addition of numerous exotic species collected from foreign places such as Asia and the Americas. Colt, a knowledgeable plantsman, added many of the *Rhododendron* genus to the garden, some of which were later replaced by better selections. He also planted numerous trees, including beeches, maples, chestnuts, and lindens. Colt also removed several buildings that he considered extraneous and intrusive to the natural environment. The Turkish Tent, the Venetian Seat, and the Chinese Alcove were all eliminated to simplify his grandfather's original plan.

The estate passed on to Colt Hoare's half brother, Henry Hugh, in 1838 and then to Hoare's son, Hugh Richard, in 1845. Hoare's nephew, Henry Ainslie Hoare, 5th Baronet, inherited it next, with little means to maintain it. He eventually abandoned it in 1886, returning only occasionally to visit the grounds. In 1894, his son, Sir Henry Hoare, 6th Baronet, inherited the estate. The landscape garden, once the pride of the Hoare family, was now in severely poor condition after nearly a decade of neglect. Sir Henry restored the garden, replacing many of the trees and ornamental shrubs. Without an heir, he decided to bequeath Stourhead to the National Trust in 1936. The Trust has since maintained the spirit of this living, artful creation that has survived over two and a half centuries.

Chapter 14 Landscape Expression: The English Landscape Park

The great landscape parks that characterized the eighteenth-century English landscape have been regarded as the product of a steady progression toward a more natural approach to garden design. This trend began with the written encouragement of Addison, Pope, and Switzer and continued with the creative genius of Bridgeman and Kent until the arrival of Lancelot "Capability" Brown, considered the master designer of the English landscape park. Garden owners, makers, and writers in the first half of the century broke tradition and became leaders in a newly developing philosophy of design. But in the second half of the century, this new tradition was developed even further with Brown's arrival on the gardening scene. He pursued an even greater naturalness in his work. His gardens, which became the most popular across England, were made to look as though they had occurred without any human intervention.

Brown (1715–1783) was born to a family of yeoman farmers in Northumberland, England. He grew up with an interest in gardening and was hired for a few small commissions, but it was his appointment as head gardener at Stowe in 1741 that would begin to define his career. His position introduced him to a number of wealthy and influential people who ultimately would become his clients. Brown established his own firm in 1750, working as an architect and a landscape designer, a career that prospered for 30 years. He became known for his expertise in visualizing the "capabilities" of a site, a term he often used to describe the site's hidden potential, and he was thus given the name "Capability" Brown. Brown had a marvelous skill for simplifying formal gardens and for creating natural-looking, scenic, and peaceful landscape settings for just about any site. He idealized nature. But his landscapes were not designed simply to provide aesthetic satisfaction to their owners, as is often suggested. His commissions owed a great deal more to economic and social factors. The landscape park was designed not only as a beautiful setting for one's country home but also to provide recreation, sport, and income. The parks grew from the lifestyle of their owners, typically the landed classes.

Landowners were savvy businesspeople, and they recognized that cash flow and the economic value of their landholdings were very important to their continued success. Landscaped parks that could produce an income meant better economic use of their land and added revenue that they could spend on other luxuries that continued to raise their standard of living (Quest-Ritson 2001). In designing a park or improving the land, it was the designer's job—in many instances, Brown's charge—to maximize the income potential of the property while making it seem as though the property had been laid out simply for pleasure. Any visual sign of "business" was removed. Farm buildings and equipment were hidden by plantings, and while grazing animals were allowed, even encouraged, to be in view, their movement was carefully controlled by enclosures and ditches, or ha-has. The landscaped parks consisted of rolling meadows and grasslands that provided handsome rents from pasture. The larger the park Brown designed into the landscape, the more income it earned for the landowner. It was therefore partly the economic value of park grazing that made the rolling green parklands in Brown's landscapes an attractive feature.

But pasture was not the only commodity derived from the park. Equally important were the clumps of trees and the shelter belts planted in the landscape. These stands of trees were selectively thinned and replanted, generating income from the timber. The woodland plantings also provided cover for game. Toward the latter part of the eighteenth century, hunting had become a popular sport of the wealthy classes, and landscape parks provided the perfect hunting ground. Lakes also were a handsome addition to the parks, used for recreation and entertainment. Both boating and fishing were a means of entertainment for invited guests.

The landscaped estate of the second half of the eighteenth century embraced nature in a very different way than it previously had. Gardens were designed to emphasize nature's beauty but also to produce income and provide opportunities for recreation. As Quest-Ritson notes, the great landscape parks were associated with a lifestyle to which all aspired. They were a symbol of great wealth and status (Quest-Ritson 2001).

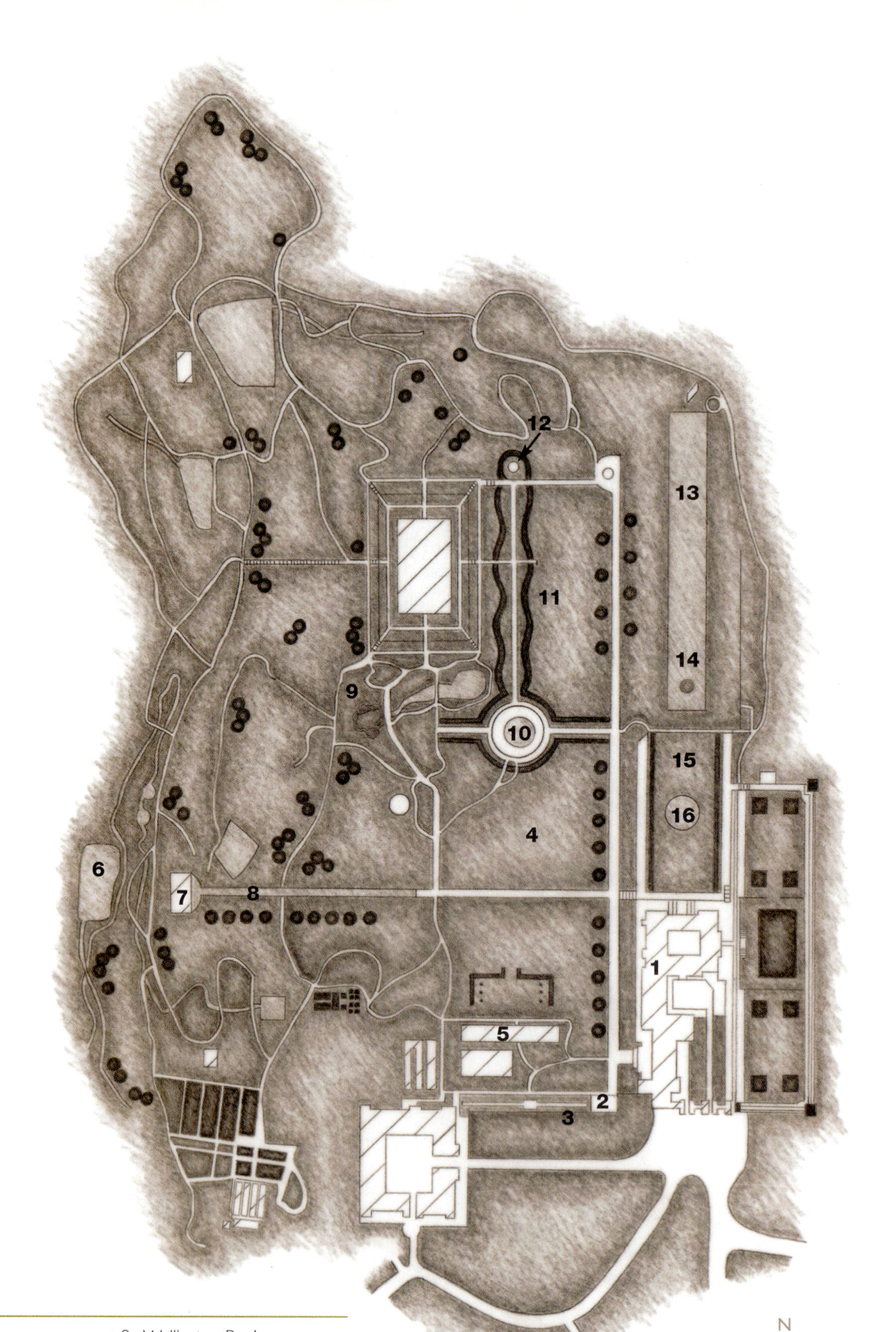

Chatsworth

1. Chatsworth House
2. Flora's Temple
3. Conservative Wall
4. Salisbury Lawns
5. 1st Duke's Greenhouse
6. Cascade Pond
7. Cascade House
8. The Cascade
9. Wellington Rock
10. Ring Pond
11. Serpentine Hedge
12. Bust of 6th Duke
13. Canal Pond
14. Emporer Fountain
15. South Lawn
16. Seahorse Fountain

Chatsworth

Derbyshire, England

Chatsworth is the product of five centuries of English garden making. Its gardens embody a rich history involving owners, gardeners, and architects who have contributed to the estate in some way.

The Chatsworth story begins with Sir William Cavendish (1505–1557), from the Cavendish in Suffolk and his wife, Elizabeth Barley (1527–1608), also known as "Bess of Hardwick." Sir William was one of King Henry VIII's commissioners during the dissolution of the monasteries. In exchange for his service to the king, he acquired land that was formerly held by the monasteries. Shortly after he married, William sold his land and in 1549 he and his wife purchased the Chatsworth property. The estate lies along the river Derwent in the Peak district, southwest of Sheffield. The gardens that encircle the house cover approximately 105 acres and are surrounded by 1,000 acres of parkland. Both the house and garden are a product of multiple layers of history, a hybrid of various styles dictated by changes in taste and economy.

In 1560, eight years after Sir William and his wife started building a house on the property, a garden was begun. Bess of Hardwick is said to have ordered herbs and flowers to be sown on the grounds. Roses, honeysuckles,

Chatsworth Time Line

1549	Sir William Cavendish and his wife, Elizabeth Barley, also known as "Bess of Hardwick," purchase the Chatsworth estate.
1560	The first gardens are created by Bess of Hardwick at Chatsworth.
1690	George London and Henry Wise are hired by Sir William, the 1st Duke of Devonshire, to design gardens with a "classic look."
1701–1703	The Cascade, originally built in 1696 for the 1st Duke, is redesigned by Monieur Grillet, a French hydraulics engineer. The Cascade House by Thomas Archer is added to the garden.
1760	Lancelot "Capability" Brown is commissioned by the 4th Duke, William Cavendish, to redesign the garden in the "Brownian" landscape park style developing in England.
1826	Joseph Paxton is appointed head gardener by the 6th Duke of Devonshire.
1830s	Hundreds of plants from other parts of the world are purchased for the garden.
1836–1841	The Great Conservatory, a huge greenhouse covering nearly one acre, was built by Joseph Paxton and Decimus Burton.
1908	The 9th Duke institutes an entry fee of one shilling for garden visits.
1920	The Great Conservatory is torn down.
1939–1946	Chatsworth becomes home to Penrhos College, an all-girls school.
1959	The 11th Duke of Devonshire and his wife, the Honorable Deborah Mitford, take up residence at Chatsworth.
1980	The Chatsworth House Trust is established to provide long-term preservation of Chatsworth for the benefit of the public.
2004	The 11th Duke of Devonshire dies, His son, the Marquess of Hartington, becomes the 12th Duke.

irises, and pansies were planted. Fruit trees were set out in rows, and garden pavilions were constructed with inspiration from Italian models. The estate passed to William Cavendish (1552–1625), the 1st Earl of Devonshire, and then to his son and his grandson, respectively.

In 1684, William, the 4th Earl, was created 1st Duke (1640–1707) of Devonshire by William III. He was recognized for his support to the king, beginning with William's ascension to the throne in 1689. The first duke had a passion for building, and shortly after his arrival to Chatsworth, he set out to make improvements to the house, changing the architectural style to give it a more classical look. In 1690, he commissioned gardeners George London and Henry Wise to design gardens that would be sympathetic to this new style. London (d. 1714) completed the west *parterre* garden and together with Wise (1653–1738) laid out the new South or Great *Parterre* Garden. Both of these men were experienced. London, who became master gardener to William III, had gained valuable insight into the methods of classical design through his study of gardens in France. Wise also was a very talented designer and horticulturist who later became royal gardener to Queen Anne and King George I. The new gardens extended the geometric forms, straight lines, and symmetrically organized spaces of the house into the garden. It was a new style to England, but one that was already well established in France and Italy, especially with the wealthy and culturally refined.

Another classical feature created for the 1st Duke's garden was the Cascade located east of the house. It was designed by Monieur Grillet, a French hydraulics engineer who had worked under the famous André Le Nôtre, royal gardener for King Louis XIV. Grillet designed the Cascade as a series of 24 stone steps that was likely inspired by Italian models. The water is supplied from a nine-acre reservoir situated just above the house. Thomas Archer (1668–1743) designed the Cascade House (c. 1701) from which the series of steps begins (Figure 14–1).

In 1702, the Canal Pond, over 300 yards long, was dug where a large hill called the "great slope" once stood at the end of the south *parterre*. The

FIGURE 14–1 The Cascade is one of the garden's main classical features.

hill restricted the view from the house across the valley toward Hardwick, another seat of the 1st Duke. Leveling this area opened the view southward, over the valley. The canal complemented the classical forms created in the garden. All of these improvements by the 1st Duke were completed just before his death in 1707.

Chatsworth was left essentially unchanged for about 60 years. It was not until the arrival of the 4th Duke of Devonshire (1720–1764) that new work began on the estate. The 4th Duke, William Cavendish, was a Whig politician and prime minister of England. In 1760, he commissioned the much-sought-after Lancelot "Capability" Brown to redesign the garden in the "Brownian landscape park" style that had grown popular across England.

Brown removed most of the 1st Duke's formal garden, sparing little except the Cascade and Canal Pond. He laid out the new Park Drive along the estate's western boundary, which was perhaps one of his greatest contributions, as it offers a variety of pleasing views of the landscape along its winding length. In 1762, architect James Paine (c. 1716–1789) constructed a new bridge that visitors arriving from Beely would travel across after making their way through Brown's panoramic landscape.

Of all of the changes accomplished by Brown, the most drastic ones, marking the shift from a formal interpretation of design to a more natural style, occurred at the series of terraces to the east of the house. The terraces, each with their own *parterre* garden, were designed along the slope of the eastern approach to the house. Brown removed the terraces, replacing them with a rolling prairie of lawn, dotted with clumps of trees that foreground a view to the wooded hillside. The 4th Duke was in charge of Chatsworth for only nine years, but the changes that he made to the garden were quite substantial and have remained much as he left them in 1764.

William Cavendish (1748–1811), the 5th Duke, did little to change Chatsworth. He died in 1811, leaving the estate to his only son, William Spencer Cavendish, who became the 6th Duke of Devonshire. The 6th Duke was a charming man in both looks and spirit. He never married and thus was known as the "Bachelor Duke." His devotion was to life's pleasures and to the peace and happiness that his home and garden provided. He made extensive improvements to both. Architect Jeffrey Wyatteville, later Sir Jeffrey Wyatteville, was commissioned by the duke to enlarge the house by adding a north wing and later an orangery. The gardens were improved slowly at first, but upon the hiring of Joseph Paxton in 1826, the duke embarked on his life's work of upgrading them.

In 1823, Joseph Paxton was a student of the **Royal Horticultural Society** (founded in 1804, it is the world's leading horticulture organization committed to educating gardeners through inspirational shows, events lectures, and demonstrations) working in the garden at Chiswick, an estate belonging to the Earl of Burlington, which adjoined the 3rd Duke of Devonshire's land. The duke came to know Paxton, and instantly recognizing his talents, he was able to recruit him for Chatsworth. In 1826, Paxton became the duke's head gardener, and together they would begin a lifetime of improvements to the gardens. For 30 years they made changes. They also

FIGURE 14–2 Tender plants such as peaches and figs, unsuited to England's climate, were grown within the glazed walls of Paxton's "Conservative Wall", completed in 1848.

broadened their gardening interests to include plant collecting, and Chatsworth became home to an extensive collection of exotic plants from around the world (Figure 14–2).

Paxton made several contributions to the gardens during his tenure, but one of his best known is the Emperor Fountain in the Canal Pond. It was built to celebrate an intended visit to Chatsworth by the czar of Russia in 1844. The czar never came, but the fountain, with water jets rising over 200 feet, was dedicated in his name and is believed to be the highest gravity-fed fountain in the world. Another of Paxton's contributions is the great rock garden that he created in 1842, a massive collection of large boulders and a large artificial waterfall, called the "Wellington Rock." Growing among the rocks are plants like mosses, liverworts, and lichens.

As a result of Paxton's passion for gardening and the duke's positive working relationship with his head gardener, Chatsworth continued to grow and prosper. In 1829, eight acres of land at the south end of the garden were set aside for the development of a pinetum, or pine garden. Six years later, an arboretum covering nearly 40 acres was established with a botanical classification system that arranged an initial collection of nearly 1,700 species by plant families. From 1830, hundreds of plants, including hardwood trees, fruit trees, flowering plants, and exotic plants from other parts of the world, were purchased for the garden. Plant-hunting expeditions to various parts of the world brought many rare and unusual species back to add to the vast collection. The growing number of exotic plants collected, especially those from the tropics, required a controlled climate in which to grow them. This led to the construction of several glasshouses. Among the most famous was the Great Conservatory, a huge greenhouse made of glass, wood, and iron, which covered nearly an acre of ground. It was built between 1836 and 1841 by Paxton and Decimus Burton, who later designed the Great Palm House at Kew. Unfortunately, the Great Conservatory no longer exists. During

World War I, the estate could not afford to heat it, thus it was shut down. The building fell into disrepair and was finally torn down in 1920, a sad moment in the history of the Chatsworth garden.

The garden changed little during the years of the 7th and 8th Dukes. In 1908, Victor Christian William became the 9th Duke of Devonshire. The 9th Duke was the first to charge an entry fee into the gardens. The charge of one shilling was originally intended to support the local hospitals, but later a portion of the income would be taken to help maintain the estate and the grounds. It also was Victor who, faced with the trials of war and the troubled state of England's economy, gave the order to demolish the Great Conservatory, one of the many nineteenth-century extravagances of Chatsworth that was much too costly to maintain. On the large plot once dedicated to the conservatory is a maze constructed with six-foot-tall yew hedges. It was planted in 1962, replacing two tennis courts that had occupied the space since the 1920s.

Edward William Spencer (1895–1950), the 10th Duke of Devonshire, and his wife, Lady Mary Cecil, moved to Chatsworth in 1939. They spent only a short time there, as war was declared the same year, and they decided to return to their home at Churchdale (Churchdale Hall). Chatsworth became home to a girl's school, "Penrhos College," until 1946. After the war, the Devonshires decided not to go back to live at Chatsworth, but they continued to maintain it through the appointment of Bert Link as head gardener. The 10th Duke died in 1950, and his surviving son, Andrew, became his heir and succeeded to the title of 11th Duke of Devonshire in 1950. He and his wife, the Honorable Deborah Mitford, moved in 1959 to Chatsworth, where the Duchess still resides. The Duke passed away in May 2004.

The 11th Duke and Duchess made several changes to the gardens. One of their first projects was to develop a more formal setting for the south front of the house. This was accomplished by framing the south lawn with double rows of tightly clipped, red-twigged lindens (Figure 14–3).

Another major work was the planting of the Serpentine Hedge (c. 1953) to create a more attractive approach to the nineteenth-century bronze statue

FIGURE 14–3 Double rows of red-twigged lindens frame the south lawn, giving the garden a very classic appearance.

FIGURE 14–4 The nineteenth-century bronze statue of the 6th Duke, by Thomas Campbell.

of the 6th Duke by Thomas Campbell, located southwest of the old conservatory garden and maze (Figure 14–4). One thousand five hundred beeches make up the green walls that lead to the 6th Duke's bronze bust, which is set high upon a pedestal at the end of the path. The hedge continues around the adjacent Ring Pond. Twelve stone busts on tapered columns designed by William Kent for Chiswick House stand against the hedge. Many of the sculptures that ornament the garden were relocated from other Cavendish properties.

Other additions in the garden by the 11th Duke and Duchess included several glasshouses, planting beds featuring new plant introductions, the maze at the site of the old conservatory, numerous pieces of contemporary sculpture, and nearly three acres of productive kitchen gardens on the site of the old paddocks.

The gardens at Chatsworth document many different periods in the history of English gardening. No one person is responsible for its layout, but each has contributed a significant piece of history to its landscape.

Blenheim Palace
Woodstock, England

A few miles to the north of Oxford lies a quaint little town named Woodstock. It is home to the magnificent eighteenth-century Churchill palace, called "Blenheim." At the end of a quiet street in town is a tall stone wall. Beyond the wall, through a monumental archway, the view opens to what Randolph Spencer Churchill (father of the great Winston Churchill) called "the finest view in England."[1] This grand estate has been home to the Churchill family for over 300 years.

The Woodstock entrance to the estate opens to a view of a vast green park with a grand lake in the foreground. A Roman-style bridge spans the lake, its majestic arches reflected in the water. Standing proudly on the horizon is the great palace of the Churchill family, built from 1705 to 1722 by the 1st Duke of Marlborough, John Churchill. Churchill acquired the property on February 17, 1705, when Queen Anne (1665–1714) presented it to him as a gift to honor his military victory against France at Blenheim, Germany. Churchill had defended Vienna from invasion by the French, defeating an army of 56,000 men assembled by King Louis XIV. It was shortly after this victory, the first against France in over 50 years, that the grateful queen granted to the Duke of Marlborough the royal manor of Woodstock. She added a sufficient amount of money, approximately 240,000 pounds, to the gift to provide for the design and construction of a fitting home for the war hero and his family. The home would be called "Blenheim."

The person chosen to design the Duke of Marlborough's palace was John Vanbrugh, a self-taught architect and former playwright. The estate was designed as a trophy, a memento of the successes achieved by a fine British military led by a devoted hero on behalf of his queen. The work on the mansion began in June 1705. The duke's wife, Duchess Sarah (1660–1744), took charge of the construction, which evolved as a series of controversies. The duchess, a most difficult woman, argued with Vanbrugh and the workmen nearly every day, forcing her opinions on them and directing substantial changes to the work they had already accomplished. The conflicts escalated, and ultimately Vanbrugh resigned from his position. In his letter of resignation, he wrote: "You have your end Madam, for I will never trouble you more unless the Duke of Marlborough recovers [from illness] so far [as] to shelter me from such intolerable treatment" (Bond and Tiller 1997).

Blenheim Time Line

1705–1722	Queen Anne presents the Woodstock property to John Churchill, the 1st Duke of Marlborough, after his victory over the French at Blenheim (Germany). Sir John Vanbrugh and Nicholas Hawksmoor begin work on Blenheim Palace. Henry Wise, Queen Anne's master gardener, lays out the gardens.
1722	The 1st Duke of Marlborough dies.
1762	The 4th Duke, George Spencer, hires Lancelot "Capability" Brown to make changes to the garden.
1817	George Spencer Churchill, the 5th Duke of Marlborough, creates a botanical garden to display exotic plantings arriving to England from plant expeditions.
1890–1935	The 9th Duke, Charles Richard John Spencer, hires French landscape architect Achille Duchêne to design an Italian garden and French-style water terraces. The duke adds nearly 500,000 trees to the park.
1986	The 11th Duke of Marlborough creates the Blenheim Foundation to help fund Blenheim's long-term management plan.
1988	Blenheim is designated a World Heritage site, only one of four such sites in the United Kingdom.
1996	Restoration plans are made and approved. Proposed improvements include selective removal and replanting of trees, as well as maintenance and repair of constructed features including the Grand Bridge, the Triumphal Arch, the Lake, and the Cascade (dam).

1. In *Blenheim Palace,* David Green (orig. 1950) Norwich, Great Britain. His Grace the Duke of Marlborough and Jarrold Publishing, 2000.

Blenheim Palace

1. Grand Avenue
2. Column of Victory
3. The Lake
4. Vanbrugh's Grand Bridge
5. Queen Elizabeth's Island
6. Blenheim Palace
7. Temple of Diana
8. Temple of Flora
9. Cascade

In the end, Blenheim came to be known as a palace of excesses. Seventeenth-century poet Alexander Pope called it "the most inhospitable thing imaginable and the most selfish" (Conniff 2000). Writer Horace Walpole described it as "execrable within, without, and almost all around" (Conniff). Intended as a high honor, Blenheim instead put the Churchill family's character in question, as its egotistic excesses brought about envy, which led to several trumped-up charges against it. Queen Anne eventually dismissed the duke from his offices, and he and the duchess fled the country in exile for two years. Upon their return and the death of Anne, the new King George I in 1719 restored the duke to his old positions. It took nearly 300 years for the Churchills to regain their honorable reputation by producing a subject worthy of such an extraordinary residence. That person was Sir Winston Churchill, one of England's greatest military heroes, born to Randolph Henry Spencer-Churchill at Blenheim in 1874.

FIGURE 14–5 Blenheim Palace, designed by Sir John Vanbrugh for John Churchill, the 1st Duke of Marlborough.

The mansion designed by Vanbrugh was of formal character, and his recommendations for the gardens followed in the same taste (Figure 14–5). The main avenues and walks were organized along strict geometries and the gardens were symmetrically balanced, along axes formed by straight avenues and garden paths. The landscape provided an effective foreground to the house. It also provided handsome views out from within the mansion. The original garden covered nearly 80 acres. It is believed to have been the work of Henry Wise, Queen Anne's master gardener, with input from both the Duke of Marlborough and Vanbrugh. The main façade of the house faces north toward the Glyme valley, which cuts through the site with a narrow stream. Vanbrugh recommended a formal approach along a long avenue that would offer several different views of the mansion along its length. This approach, called the Grand Avenue, was made to cross the Glyme valley by one of Vanbrugh's most significant contributions to the garden, the Grand Bridge,

FIGURE 14–6 One of the finest features of the garden is the Grand Bridge, by John Vanbrugh.

a monumental structure that he considered the "finest bridge in Europe" (Bond and Tiller 1997) (Figure 14–6).

The south gardens were divided into separate areas that included a walled kitchen garden, a flower garden, and an ornamental *parterre* that included several fountains. On the south side of the *parterre* garden was a planned "wilderness," called the "Woodwork," which contained a number of different tree species as well as formally clipped ornamental shrubs and evergreens.

The Duke of Marlborough died in 1722, but the duchess remained at Blenheim and continued to make changes to the garden. In 1723, she recalled Nicholas Hawksmoor to the garden to erect the Triumphal Arch to honor her husband. Several years later, she commissioned masons Townscend and Peisley to construct the Column of Victory, a simplified version of Hawksmoor's original proposal. This **Doric** (a style of architecture characterized by a fluted columnar shaft and no base) column, 134 feet high, was positioned at the entrance to the Grand Avenue, situated along the skyline when viewed from the north side of the palace. At the top of the column is a lead statue of the duke, which stands proudly in celebration of his triumphs.

Lancelot "Capability" Brown, the highly sought-after landscape designer of the eighteenth century, was called to Blenheim in 1762, at the peak of his career, by the 4th Duke, George Spencer (1739–1817). Brown's work was the popular fashion among the social elite, and his advice for improvements to landscaped grounds was much desired during the second half of the eighteenth century. Brown's greatest contribution to Blenheim was his transformation of the park into a united composition. Vanbrugh, Hawksmoor, and Wise had created several unique and interesting elements in the garden—a spectacular bridge, an arch, and a column, as well as a series of well-designed formal gardens—but as wonderful as these features were, they did not seem harmonious. Brown's work blended the garden's elements into a seamless composition of individual features.

These individual elements were brought together in a natural, park-like landscape that covered nearly 2,500 acres. Brown carefully composed the views, improving sight lines and denying vistas until he deemed them most advantageous from a particular vantage point. Trees were planted around the site, arranged in clumps, shelterbelts, and woodlands. The clumplike plantings shielded or delayed views, the shelterbelt plantings that surrounded the park gave the landscape a much-needed sense of enclosure, and the woodland plantings provided cover for hunted game as well as a potential source of income as timber.

One of Brown's most significant improvements to the park was the Great Lake, the largest he ever designed. It covers nearly 150 acres, appropriate to the scale of Vanbrugh's palace and bridge. The lake follows the meandering Glyme valley, extending from the northeast to the southwest of the area adjacent to the palace on the north side. Through careful engineering, Brown was able to adjust the lake's water level. He raised it 15 feet to give purpose to Vanbrugh's monumental bridge, an improvement to what Vanbrugh had envisioned.

From the Great Lake, the Glyme falls over a dam disguised as a natural-looking rocky wall through which white water rapidly flows. This creation, called the Cascade, was designed by Brown in the 1760s (Figure 14–7). On the adjacent land below and around the Cascade, George Spencer Churchill, the 5th Duke of Marlborough (1766–1840), created a botanical garden displaying many of the new varieties of plantings arriving into England during the nineteenth-century fad for plant collecting.

The 9th Duke, Charles Richard John Spencer Churchill (1871–1934), made several additions to the garden between 1890 and 1935. He hired French landscape architect Achille Duchêne to design an elegant, formal garden in the Italian style. This garden, on the site of the 1st Duchess' flower garden, is designed as a *parterre* with evergreen topiary and an elegant fountain. The patterned beds are punctuated with potted citrus from the nearby orangery during the summer months. Duchêne also designed the magnificent French-style water terraces on the west side of the palace (Figure 14–8). These terraces were inspired by the great water gardens designed by André Le Nôtre at Versailles. The duke also added several hundred thousand trees to the park, including several non-native species, which would not have been acceptable to Brown. The blue atlas cedars and the copper beeches, for example neither of which is native to England, would likely never have been permitted in one of Brown's "natural" landscapes.

The current owner, John George Vanderbuilt Henry Spencer-Churchill, the 11th Duke of Marlborough, is responsible for Blenheim's long-term management plan that provides for continuous replacements and enhancements in the gardens. This work is funded partly by the Blenheim Foundation,

FIGURE 14–7 The Cascade, designed by Lancelot "Capability" Brown, was an aesthetic solution to disguise the overflow point of the dam created for his 150-acre lake.

FIGURE 14–8 The water terraces on the west side of the palace were developed by French landscape architect Achille Duchêne. His vision was inspired by the accomplishments of the great André Le Nôtre.

which also was created by the 11th Duke and family trustees in 1986. The goal of the foundation is to raise money to provide for the maintenance of the palace and grounds.

The great palace of Blenheim has been made famous through the lives of two of its most noteworthy residents, the 1st Duke of Marlborough and Sir Winston Churchill, but it has also gained equal recognition for its grand architecture and impressive landscape. It was designated in 1988 a World Heritage site, only one of four sites chosen in the United Kingdom. To be considered for the World Heritage list of protected properties, a site must possess universal value of a cultural or natural significance. Blenheim meets that criterion. Today it is regarded as one of the most internationally famous works of landscape art, rich in historical associations, with several layers of English landscape garden history contained in one fascinating site.

Chapter 15

Landscape Expression: The Nineteenth-Century Victorian Garden

The Victorian period in England coincides with the reign of Queen Victoria, from 1837 to 1901. From 1815, the date marking the end of a long war between Britain and France, England was a prosperous country with strong and steady economic growth as a result of the **Industrial Revolution** (a major change in manufacturing beginning in England during the eighteenth century whereby human labor was replaced by machines). Nearly every segment of society was able to benefit from the great economic changes. For the aggressive and self-motivated individual, wealth was no longer a fantasy; it was accessible, regardless of birthright.

A newly rich class of people emerged, having made their money in manufacturing and working as merchants. With their newly acquired wealth, these people began to seek recognition among society's most prestigious groups. Education, proper etiquette, and taste were all part of the qualities belonging to high or so-called "polite" society. They were the qualities of a "gentleman." Many of these socially acceptable characteristics often took several generations for a newcomer to high society to acquire, while other requirements, such as property ownership, were fulfilled rather quickly. Thus this new wealthy class began to acquire land upon which extravagant homes and gardens were built to exploit its wealth. While these homes were impressive in scale, they were less so in appearance. Victorian architecture was cloaked in many different period styles, though most lacked a meaningful context. Victorian gardens were similarly incoherent. The idea of having a garden appealed to the newly rich landowners, as gardens were regarded as a socially acceptable pursuit of the wealthy. But their gardens, characteristically extravagant and grand in scale, were like the period architecture, a combination of many different styles copied and often inharmoniously combined.

The new trends in gardening during the Victorian period were underpinned by many factors, including changes in taste, advances in technology, and a growing passion for horticulture. The greatest influence on the development of gardens, however, was new wealth. Victorian garden design became a celebration of wealth and a search for aesthetic fulfillment, inspired by the classic gardens of Italy and France. Elements of classic design, such as colorful *parterres,* fountains, and sculptures, were combined with other period elements and traditions from around the world. Together these elements created an exceedingly eclectic framework, often quite showy and inharmonious.

One unique aspect of Victorian gardening was the inclusion of an extensive variety of "exotic" plants in the garden. Toward the end of the eighteenth century and well into the nineteenth century, nurserymen and well-to-do amateurs alike developed a passion for collecting and cultivating exotic plants from the far reaches of the Pacific, China, South Africa, and the Americas. These plants found their way into gardens that were designed especially to help accommodate the growing requirements of species arriving from all over the world. Advances in technology helped fuel this trend. In 1833, Nathaniel Ward invented the Wardian Case, a miniature, portable greenhouse that enabled live plants to be transported over long distances. Without it, the number of plants arriving to England would not have been anywhere near what it was.

The development of glasshouses in England also was a major factor in the growing trend of plant collecting. The earliest glasshouses were built to protect plants from frost damage during the winter months. They were unheated at first, but by 1650, charcoal burning stoves were installed in them as a source of heat. By the 1660s, glasshouses were constructed with heated walls that protected the most tender of plants. By the 1830s, this method was replaced by hot water systems that proved to be more efficient. One of the greatest examples of this technology was Joseph Paxton's Great Conservatory at Chatsworth Garden in Derbyshire. This structure, the largest of its kind in the world at the time, gained Paxton international notoriety. The glasshouse, covering nearly an acre, housed everything from the smallest aquatic plants to large palm trees.

Another invention tied to advances in technology was the lawn mower, patented in 1830 by Englishman Edwin Budding. This gained immediate popularity and had a considerable impact on garden aesthetics. The mower gave lawns a new, manicured look that became highly desirable in gardens. Lawn mowing also was considered an acceptable "gentleman's" activity. It provided exercise and a way to become involved in some sort of gardening activity.

New industrialized printing technology also impacted garden design. By the 1840s, advanced systems for commercial printing permitted the publication of garden books, periodicals, and mail-order nursery catalogues. New plant cultivars were made known and available to the public throughout the country. Journals such as *The Gardener's Gazette, The Horticultural Magazine, The Cottage Gardener,* and *The Gardener's Magazine* were just some of the many garden periodicals published in the 1840s. These new magazines provided horticultural information, book reviews, plant introductions, and critical reviews of popular gardens. Commercial printing also gave nursery companies an opportunity to advertise their products and sell them by mail order to a vast number of potential consumers.

The Victorian period, though most remembered for its excesses and questionable taste, made a significant contribution to the development of the landscape with the creation of public parks. One of the main goals for the establishment of these parks was to give people living in the city an opportunity to escape their crowded living environments. The population had more than doubled in the nineteenth century, and the cities were saturated with houses and factories with little open space. The first public park to be built that permitted free entry was Birkenwood in Liverpool. Designed by Joseph Paxton in 1843, it was a proven success and a model for future parks. Central Park in New York City, for example, designed in 1857 by Frederick Law Olmsted (1822–1903) and Calvert Vaux (1824–1895), was inspired by Birkenwood. By the end of the nineteenth century, most cities throughout England had a public park.

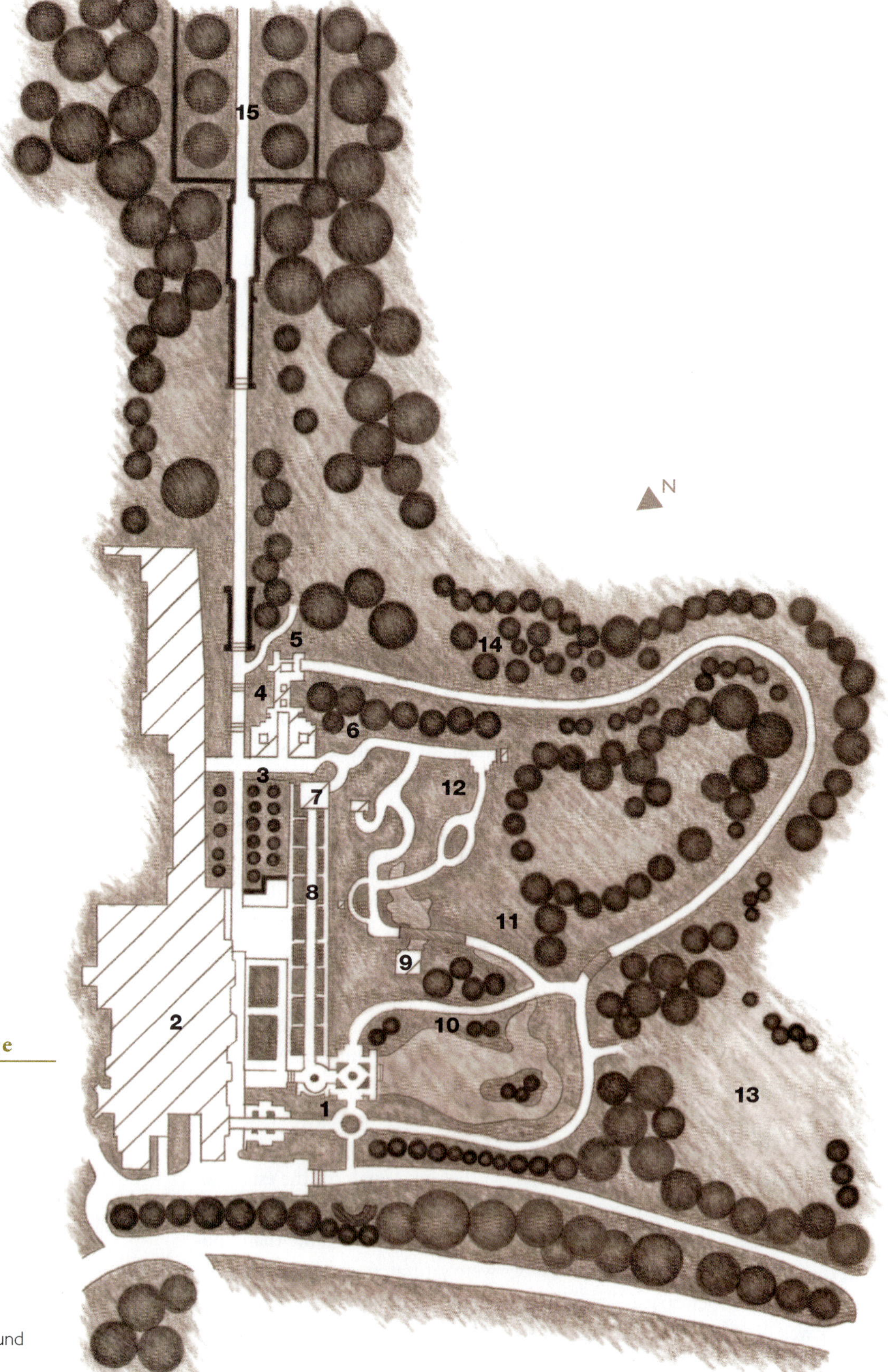

Biddulph Grange

1. The Italian Garden
2. House
3. Cherry Orchard
4. Egypt
5. Cheshire Cottage
6. Stumpery
7. Shelter House
8. Dahlia Walk
9. Temple
10. Rhododendron Ground
11. Glen
12. China
13. Tennis Court Lawn
14. Pinetum
15. Wellingtonia Avenue

Biddulph Grange
STAFFORDSHIRE, ENGLAND

The nineteenth century was an important period in English garden history, not in terms of its contribution to garden design but rather as a period of significant growth and advancement in the horticultural sciences. There was a growing passion for plants and plant collecting as well as an interest in horticultural innovation and invention. Plant collecting and breeding had become quite popular with all segments of society, especially the wealthy classes. Many gardens built during this period were designed to accommodate these collections. Biddulph Grange is one such garden, the creation of three avid gardeners: James Bateman and his wife, Maria, its owners, and their friend, Edward Cooke, a painter and knowledgeable plantsman. James Bateman was born into a wealthy family of industrialists. His father owned several factories that manufactured steam engines, boilers, and a wide range of other products. They also owned several cotton mills in Manchester.

As a young man, Bateman had developed a passion for horticulture. He became a collector of orchids at age eight, and by the time he entered Magdalen College in Oxford, he was sending plant collectors to foreign lands to collect new species for him. On April 24, 1838, he married Maria Egerton-Warburton, who was also an avid gardener. Two years after their marriage, the couple moved to Biddulph and added on to an old vicarage (residence of an anglican priest known as a vicar) once occupied by James' uncle, William Holt, who served as the vicar of Biddulph from 1831 to 1873.

In September 1849, the Batemans, eager to create a garden around their home, invited Edward Cooke (1811–1880) to visit Biddulph. Cooke was the son-in-law of nurseryman George Loddiges, owner of the Hackney Nursery, which at that time was said to have had the largest collection of plants in the entire world. It was through this association that young Cooke learned about plants and the practicalities of gardening.

With Cooke's help, the Batemans began to create an extraordinary garden with an extensive collection of plants from around the world. They designed the gardens as a series of picturesque scenes created as separate areas, each with its own specifically designed habitat that simulated the plants' indigenous environment. The plants were carefully selected and their location adapted to provide the perfect growing conditions—dry or moist, sunny or shady—for the plantings. Architectural features and constructed elements contributed to the picturesque scenery.

The very first gardens to be laid out were the formal *parterre* gardens at the south front of the house. These gardens, framed by stone walls and yew

Biddulph Grange Time Line

1810	James Bateman, an industrialist, purchases Biddulph Grange and Knypersley Hall.
1838	James Bateman's grandson, James, marries Maria Egerton-Warburton and takes up residence at Biddulph two years later.
1849	Bateman hires marine painter and garden enthusiast Edward Cooke to assist in designing the garden. An extensive collection of exotic plants is introduced into the garden from around the world.
1868	An elderly James Bateman passes on the estate to his eldest son, John, and his wife, the Honorable Jessy Caroline.
1872	John Bateman sells the estate to Robert Heath, an industrialist and a member of parliament for Stoke.
1921	Heath offers Biddulph Grange to the North Staffordshire Cripples Aid Society for conversion to an orthopedic hospital. The hospital opens three years later.
1926	Biddulph is sold to the Lancashire Education Committee for use as a hospital for crippled children.
1988	The National Trust acquires Biddulph Grange and embarks on a major restoration of the house and gardens.
1990	Restoration begins on Egypt.
1996	Restoration begins on the Wellington Avenue.

FIGURE 15–1 Formal terraces in the Italian garden are planted with flowers, azaleas, evergreens, and a stand of Italian cypresses.

hedges, were planted with annual flowers and roses. Below the *parterres,* on an east-west axis, the Batemans created the Dahlia Walk. Dahlias, known for their spectacular range of colors, were popular in England during the early 1800s. To the east, along the south face of the house, a cherry orchard was planted, and beyond it a small arboretum with trees such as sweet gums, hollies, and maples. The east axis continues as a long avenue, Wellingtonia Avenue, named for the giant American sequoia trees, or *Wellingtonias,* that were planted along its length on either side. Bateman, had obtained some of the earliest specimens of this impressive tree from America in 1853. He planted the trees along the avenue, alternating each with deodar cedars, which were intended for removal as the *Wellingtonias* matured. It would be Bateman's successor, Robert Heath, who would ultimately remove the majestic *Wellingtonias* trees instead.

The stepped terraces that descend the hill to the south toward the lake are referred to as the "Italian Garden" (Figure 15–1). Its reference to the Italian style is likely owed to those planted terraces. The western boundary of the garden is defined by an avenue of linden trees, or "limes," as the English call them. This planting, which predates the Bateman family, once bordered a roadway connecting Biddulph to nearby Congleton.

One of the most beautiful gardens at Biddulph is the Rhododendron Ground. These "American" plants, as they were indiscriminately called (ignoring many of their true origins), were a favorite of Bateman's and were selected for the area surrounding the lake. Cooke designed rock outcroppings to create a "natural" wooded environment, similar to the plants' native habitat.

The Egyptian garden is one of Biddulph's unexpected pleasures. Concealed behind a tall, tightly clipped hedge are stone sphinxes that guard the entrance to the space called Egypt. This garden features topiary yews clipped to form obelisks and a pyramid. Europeans developed an appreciation for Egyptian art and architecture during the early 1800s, which may have been the inspiration for this garden area.

The Cheshire Cottage is actually a brick building clad with a wooden cottage-like façade. It functions as the exit from the Egyptian garden and also creates a gateway into the **pinetum** (an arboretum of coniferous evergreens, especially pines). The pinetum displays another of Bateman's horticultural favorites—pines. It is not only a collection of pines, as the name implies, but consists of many different conifers. The trees are presented on gentle berms created to display them better and also to provide a more favorable soil environment, with good drainage. The collection includes many unique specimens, such as Douglas firs, deodar cedars, atlas cedars, Japanese cedars, and *Wellingtonias,* all introduced between 1826 and 1853.

One of the most popular gardens at Biddulph is "China" created to display Bateman's collection of plants from the Far East. Many unique architectural features designed by Cooke were added to this garden to make it look more like China. Exotic buildings of many different styles, including Chinese, were popular features in eighteenth- and nineteenth-century English landscape gardens (Figure 15–2). The Chinese temple designed for Biddulph Grange was among the finest representations of this type of architecture in Europe. The temple, with its adjoining terrace, overlooks a lake and a colorful wooden footbridge that crosses the water (Figure 15–3). *Hydrangea, acuba, camellia,* and Japanese maples were among the plants displayed in the garden. The China garden occupies a secluded location at Biddulph. It is set apart from other areas by large mounds of earth that resulted from the excavation of the lake, not only creating a dramatic effect but also providing necessary shelter for the tender plants within the garden.

FIGURE 15–2 Architectural features helped establish a Chinese theme in the China garden at Biddulph.

In 1868, an elderly James Bateman, no longer able to keep up with the costs and the responsibilities of maintaining Biddulph, surrendered it to his eldest son, John, and his wife, the Honorable Jessy Caroline. It was not long before John decided to sell the property. Robert Heath, a leading industrialist and member of parliament for Stoke in the 1870s, purchased Biddulph in 1872. Though not an avid gardener, Heath did a fine job maintaining the gardens. World War I had a devastating effect on Britain, and many successful companies suffered the consequences of the war. Robert Heath and Sons, the company owned by Heath, was no exception. His failing business forced him to part with many of his assets, including Biddulph. In 1923, Biddulph Grange was acquired by the North Staffordshire Cripples Aid Society, who converted it into an orthopedic hospital. In 1926, it was sold again. Lancashire Education

FIGURE 15–3 A colorful wooden bridge spans the lake in Biddulph's China garden.

Committee purchased Biddulph to create a hospital for crippled children. It remained a hospital for several decades, and the gardens continued to be adequately maintained when the hospital was phased out.

The **National Trust** (an organization founded in 1895 to act as guardian for the nation in the acquisition and protection of threatened countryside, including noteworthy gardens and historic buildings), recognizing the historical importance of the garden, purchased it in 1988 and shortly thereafter embarked on one of its most comprehensive garden restoration projects, returning Biddulph's garden to its original glory as one of the Victorian era's most outstanding and innovative garden accomplishments.

Chapter 16 | Landscape Expression: The English Flower Garden

By the early 1900s, the fashion of the wealthy owning large estates had faded. Properties were acquired in smaller parcels, with most of the money for land purchases earned in industry or from family inheritances. Agriculture was no longer an activity associated with most properties, thus there was no need for thousands of acres; a few hundred was sufficient for any one family to enjoy the benefits of country life. Early-twentieth-century landowners also had a different idea of the garden or landscape than their landowning predecessors. The nineteenth-century age of industrialism had produced a major change in the English landscape. Mines, factories, and mills replaced what were once acres of beautiful countryside. Society, though content with the country's economic progress brought about by industry, was devastated by the extent of destruction it caused to England's natural countryside. By the end of the century, people, reacting against this large-scale destruction, began to develop a greater appreciation for the beauty of wild, untouched nature. This new awareness led to the desire for more natural-looking gardens, a concept that had first appeared in English culture during the eighteenth century, when society reacted against the formality of seventeenth-century design.

Author and gardener William Robinson (1838–1935) led a literary campaign late in the nineteenth century directed toward encouraging a more naturalistic mode of gardening. In his work and writing, beginning with his first book, *The Wild Garden,* published in 1870, he expressed the need for a greater respect of nature in garden design. Robinson promoted the virtues of working with, rather than against, nature. He expressed the need for gardeners to recognize the natural conditions of a site and to establish plant communities suited to those conditions. His advice for observing nature as a guide to creating gardens was followed by a great number of garden designers, including English garden designer Gertrude Jekyll (1843–1932), who expanded on his theories and extended the popularity of this new style even further. Jekyll championed naturalistic plantings in the rural cottage gardens that she designed during the late nineteenth and early twentieth centuries. She took the knowledge that she gained from observing growth patterns and plant combinations in the wild and applied it to her planting designs, making them appear to have evolved from nature. Plants were arranged in naturalized drifts, with rich combinations of colors, textures, and forms. Her gardens appeared to have developed without plan, owing their beauty to splendid, artistic combinations of hardy plantings. Indeed, that was the essence of her designs.

Jekyll's skill for composition, especially her appreciation for color, was gained during her first career as an artist, which played a major role in the success of her creations. Her approach to garden making grew increasingly more popular and was soon recognized as a distinct style of design. In 1899, she and architect Edwin Lutyens (1869–1944) combined their skills in a partnership that lasted over two decades. In their gardens (which totaled over 100) Lutyen's formal geometries were softened by Jekyll's natural-style plantings. One of their most successful collaborations was Hestercombe Garden in Somerset, commissioned in 1906. There, Lutyen's architectural layout of terraces, walls, stairs, and pergolas is softened by Jekyll's luxurious combination of plant colors, forms, and textures. This sort of gardening style influenced the design of many successful gardens in England. Sissinghurst in Kent, one of England's most popular gardens, designed by Vita Sackville-West (1892–1962) and her husband, Harold Nicholson (1886–1968), in 1930 is one such example. Sissinghurst's popularity is owed to its splendid display of plant combinations designed in the manner of Robinson and Jekyll and organized within the framework of a series of garden rooms. It, too, has been an inspiration to many garden makers throughout England and in other parts of the world.

This gardening style continues to be practiced in Britain and to some extent in America. It has gained even more popularity in recent years, as many new gardeners have begun to develop more ecologically conscious designs with native plants or plants with similar characteristics and cultural requirements as those native species.

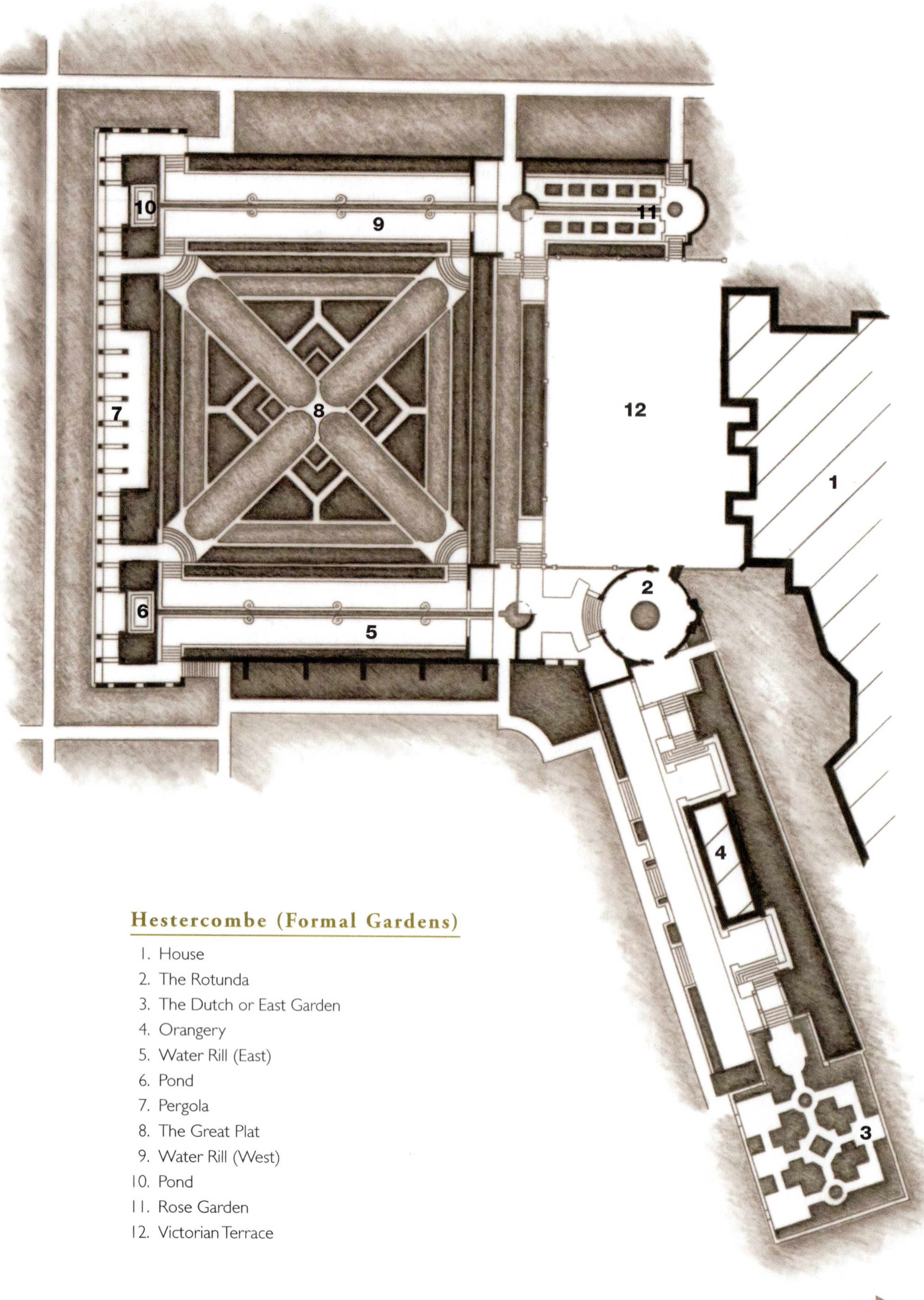

Hestercombe (Formal Gardens)

1. House
2. The Rotunda
3. The Dutch or East Garden
4. Orangery
5. Water Rill (East)
6. Pond
7. Pergola
8. The Great Plat
9. Water Rill (West)
10. Pond
11. Rose Garden
12. Victorian Terrace

Hestercombe

SOMERSET, ENGLAND

Hestercombe in Somerset, has one of the finest English gardens created during the twentieth century. It is the product of a successful collaboration between an architect and a gardener and one of the first examples of a garden that combines both formal and informal aspects of garden design. The formal gardens built between 1904 and 1908 are the most popular attraction at Hestercombe, but the earliest gardens on the estate date back to the mid-eighteenth century. Both period gardens have been recently restored to their original glory, and both are excellent representations of their respective period's manner of gardening.

Hestercombe's history has been traced back to 854, when it appeared in an Anglo-Saxon charter, through most of its history is associated with the Warre family, who owned the property for nearly five centuries, beginning in 1391 and ending in 1872 with the death of Elizabeth Warre. In 1731, a parliamentarian named John Bampfydle married the daughter of Sir Francis Warre. His son, Coplestone Warre Bampfydle, an artist, inherited the property from his father in 1750. That same year, he laid out the design for a landscape garden in a long, narrow valley running north to south on the site. The mature woods, with its natural springs, offered a perfect setting for the type of landscape Bampfydle envisioned. His plan for a picturesque landscape, complete with classical buildings, was realized along the valley's wooded hillsides. From these buildings planned views to the other garden features as well as views to the neighboring countryside could be enjoyed. The landscape garden, which covered over 35 acres, was designed by Bampfydle to be a journey filled with surprises. It is laid out as a series of planned views, calling attention to both natural and constructed features in the landscape (Figure 16–1).

FIGURE 16–1 One of the main features of the landscape garden is the Great Cascade, surrounded by woodland.

Hestercombe Time Line

- 854 Hestercombe is listed on an Anglo-Saxon charter.
- 1391–1872 The property is owned by the Warre family.
- 1750–1786 Coplestone Warre Bampfydle inherits the property from his father, John Bampfydle, and begins laying out a landscape garden.
- 1873 Hestercombe is acquired by 1st Viscount Portman, who transforms the eighteenth-century Bampfydle house to suit his tastes for nineteenth-century Victorian fashion.
- 1892 The property is given to Portman's grandson, the Honorable Edward Portman, as a wedding gift.
- 1904 Architect Sir Edward Lutyens is hired to redesign the house and gardens, along with artist and gardener Gertrude Jekyll.
- 1942 Viscount Portman dies. The house is leased to the Somerset County Council as headquarters for the County Fire Service.
- 1973 Jekyll's original landscape plans are discovered. A comprehensive garden restoration project begins.
- 1997 Bamfydle's eighteenth-century landscape gardens are restored.

The constructed elements introduced by Bampfydle were designed as "**eye-catchers**" (a feature, often architectural in nature, placed on a distant and prominent point within or visible from the garden to accent or direct a view) in the landscape. They also were destinations along the garden's route, from which carefully chosen garden prospects could be enjoyed. The recently restored Temple Arbor, built in the 1770s in a classic style, provides an extraordinary view of the pear-shaped lake, Pear Pond, from its hillside prospect. Another building, the Witch House, built in 1761, stands at the summit of the adjacent hillside. It too is constructed as a temple, but here instead of paying tribute to the virtuous, it seems to be associated with the allure of the unknown. Another building, the Mausoleum, dating from the mid-1750s, is named purely for its architectural style. It was designed simply as an artful element in the landscape, not as a funeary building. The Friendship Urn (c. 1786), one of Bampfydle's last additions to the garden, was created to memorialize two of his closest friends, Sir Charles Kemeys-Tynte of Halswell and Henry Hoare II of Stourhead in Wiltshire. The urn is a copy of an earlier-designed piece by eighteenth-century landscape designer William Kent for one of his clients, English poet Alexander Pope.

In 1873, Hestercombe was acquired by 1st Viscount Portman, who commissioned a transformation of the eighteenth-century Bampfydle house to suit his tastes for the nineteenth-century Victorian fashion. Unfortunately the nineteenth-century renovations essentially ruined what was otherwise a fairly nice looking home. In 1892, the house was given to the Honorable Edward Portman by his grandfather as a wedding gift. It was Portman who commissioned architect Edwin Lutyens in 1904 to design a new garden, which became a major feature of the estate and a significant achievement in uniting nature and art. Moreover, the garden was a pioneering accomplishment in design, a collaboration between two separate design professions. Lutyens worked together with artist and gardener Gertrude Jekyll. Their individual expertise was successfully united in this work, creating an extraordinary garden combining the formal and the informal in a complementary way.

Lutyens began his architectural career in 1888. His practice remained active for nearly 50 years, in which time he became known as one of the best, most-sought-after architects in Britain during the twentieth century. Lutyens was a traditionalist with a modern awareness whose exceptional skills as an architect were revealed particularly in his domestic accomplishments. He was responsible for the design of many great country homes throughout England, that were built or redesigned during the early twentieth century. Lutyens's attention to both interior and exterior space makes his work unique. He was a master of spatial organization and composition and of integrating building and site. He worked very hard to create interior plans that established strong connections between house and garden. For Lutyens, the idea of the garden having a strong connection to the country house was not simply a tasteful quality, it was the essence of his designs. The whole idea of a house in the country was its connection to the landscape. It was not enough to be located in a country setting. To be successful, the house needed to become a part of its environment, a natural outgrowth of the landscape. His garden spaces, like his architecture, were typically organized in a formal, symmetrical fashion. He designed his gardens as a series of outdoor rooms to be lived in as an extension of the home.

One of the most beneficial relationships in Lutyens's career as an architect was with Jekyll. They collaborated on a number of projects. Her expertise in horticulture and her keen sense of design resulted in artistically composed planting compositions that blended nicely with his classical style of architecture. Jekyll proposed naturalistic plantings as a way to soften architectural lines and features. Her artist's eye helped her form successful compositions using form, texture, and color to their best advantage. Together they established harmony between the formal and informal, two competing traditions or theories of design that were greatly at odds during the early part of the twentieth century.

Hestercombe's formal garden is one of the finest gardens designed by Jekyll and Lutyens. The largest garden area, south of the house, is called the "Great Plat" (Figure 16–2). It is designed as a sunken *parterre* garden laid out with geometric borders edged with stone paths reminiscent of the gardens of Renaissance Italy. The space is divided by two long diagonal paths crossing in the center and terminating at each corner of the garden in beautifully detailed stone steps designed as quarter circles. These steps lead to the next garden level on either side of the Great Plat. To the east and west are broad, raised

FIGURE 16–2 A sunken parterre *garden, known as the "Great Plat," is the largest of the gardens at Hestercombe.*

FIGURE 16–3 A long channel of water or rill creates an axis toward the distant horizon.

paths bisected by long channels of water, the east and west rills (Figure 16–3). Each rill begins from a small pool enclosed in grotto-like niches set into split-stone walls. These walls are elegantly draped with fruiting vines. The rills terminate in small, rectangular pools filled with water plants at the opposite end of the garden. At the southern end of the formal *parterre* is a pergola extending the entire width of the garden, forming its southern boundary. The structure, covered by a canopy of climbing roses, clematis, and honeysuckle, provides a series of framed views of the surrounding English countryside. At the north end of the east water garden, a broad flight of stone steps leads to the rotunda, a circular space surrounded by high stone walls. An interesting paving pattern created in natural stone radiates from a round pool that occupies the center of this space. The rotunda unites several of the garden's axes—the south terrace, the east water garden, from which the main garden extends to the west, and the orangery.

From the rotunda, steps lead to the orangery, a building designed by Lutyens in his trademark classic style. It is a handsome structure made of local material, a yellow Somerset stone from the Ham Hill district. It is elegantly designed with a high level of detail and ornament. The orangery was intended more as a sort of garden pavillion than it was as a place for the overwintering of plants. On the east side of the orangery, steps lead to the Dutch garden. The design of the space is simply stated, with a formal arrangement of planting beds separated by stone paths. Italian urns punctuate the space, giving the garden arrangement structure and visual interest.

Hestercombe is a significant example of the successful synthesis of a formal design approach with a naturalistic planting style. Its pioneering design accomplishments have had lasting value as a testimony to the merits of each approach.

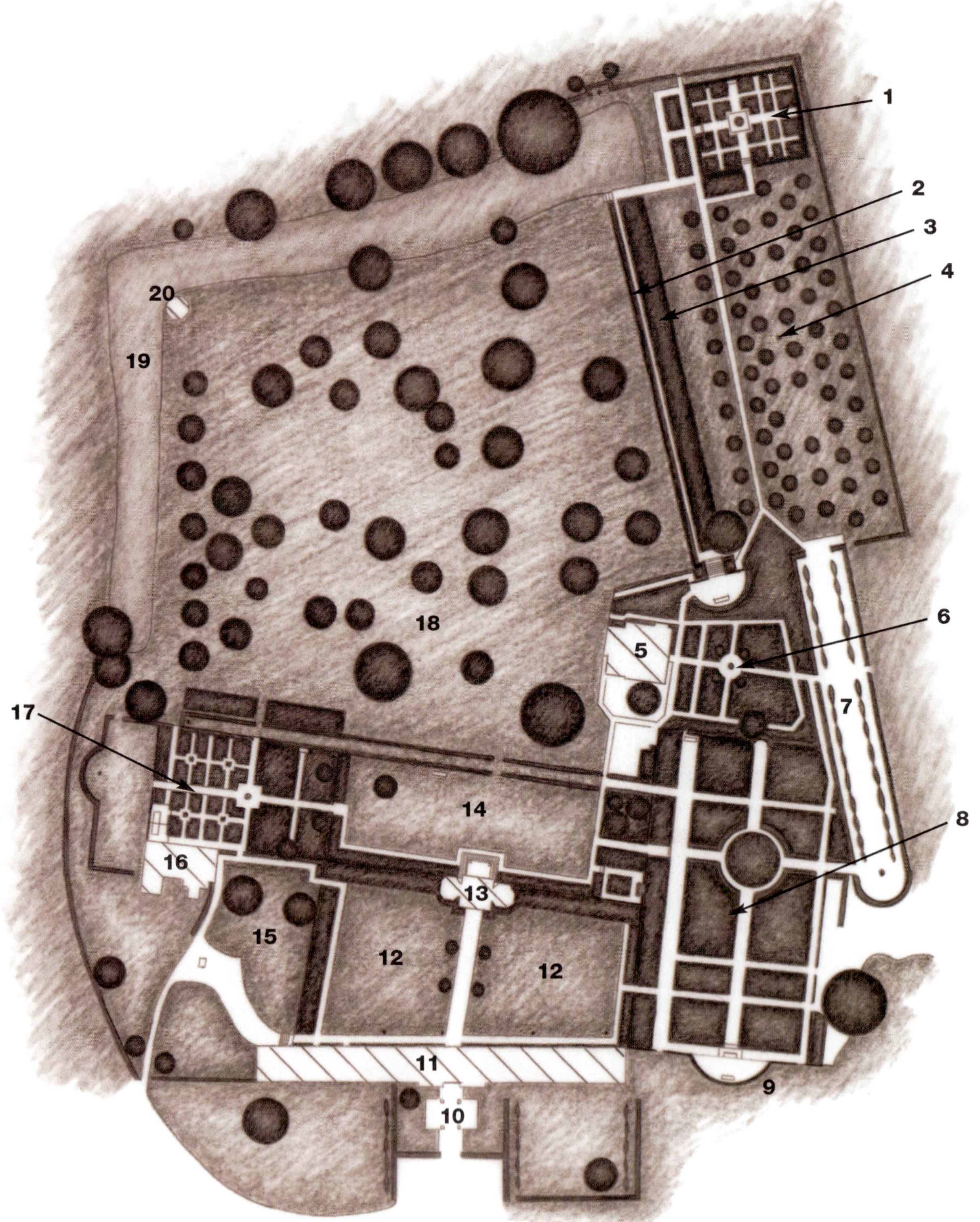

Sissinghurst Garden

1. Herb Garden
2. Moat Walk
3. Azalea Bank
4. The Nuttery
5. South Cottage
6. Cottage Garden
7. Lime Walk
8. Rose Garden
9. Powys's Wall
10. Forecourt
11. Main House
12. Front Courtyard
13. Tower
14. Tower Lawn
15. Delos
16. Priest's House
17. The White Garden
18. The Orchard
19. Moat
20. Gazebo

Sissinghurst Garden

Kent, England

Vita Sackville-West (1892–1962), a poet and novelist, and her husband, diplomat and writer Harold Nicolson (1886–1968), were passionate gardeners whose love and dedication to their hobby inspired many gardeners, both amateur and professional. Their garden, Sissinghurst, is located two miles northeast of Cranbrook in the Weald of Kent. Little is known of Sissinghurst's earliest history. It was first mentioned in written documents back in the twelfth century. The property was bought and sold numerous times through the centuries, with buildings both added and removed to suit the needs of each family that lived there.

Vita and Harold first visited Sissinghurst in April 1930. They were looking for new property in Kent, as their own beloved and very peaceful property, Long Barn, two miles from Knole, was being threatened by a proposal for a sizable poultry farm to be constructed in the fields surrounding their estate. The couple learned of the Sissinghurst property through an advertisement that described the Victorian farmhouse and six cottages with architectural ruins throughout its fields of nearly 400 acres. The Nicolsons, especially Vita, envisioned Sissinghurst's potential, and in May 1930, they purchased the property for 12,375 pounds. They moved to Sissinghurst two years later and began a major renovation of the estate, a task that dominated most of their time for the better part of five years, during which they began to establish a garden.

Sissinghurst Garden Time Line

1150–1250	The Saxinghurste family occupies the estate.
1250–1930	The property is bought and sold several times.
1565	The Tower is built.
1620s	The Priest's House is constructed to house Sissinghurst's own chaplain.
1756–1763	The estate is leased to the government for use as a prisoner of war camp during the Sevens Years' War.
1930	Vita Sackville-West and her husband, Harold Nicolson, purchase Sissinghurst.
1932–1936	The Nicolsons establish the gardens at Sissinghurst. Architect Albert Powys works as a consultant.
1939	The gardens are transformed into fields for harvesting during wartime.
1945	A restoration of the garden begins.
1962	Vita Sackville-West dies, leaving the estate to her younger son, Nigel, with life rights to the South Cottage bequeathed to her husband, Harold.
1967	The Sissinghurst property is transferred to the National Trust to avoid payment of the estate duty tax.
1968	Harold Nicolson dies.
1994	Head gardener Sarah Cook is appointed as Sissinghurst's property manager.

FIGURE 16–4 The Rondel Rose Garden. Note the curved wall designed by Albert Powys who was inspired by the green palisade at Villa Gamberaia in Settignano, Italy.

Harold commissioned the building of a dam in the lower field to make a lake. Other improvements were initiated upon the advice of an architect named Albert Powys. Though Vita always contended that she and Harold independently restored the estate, with no professional help, it must be pointed out that from 1932 to 1936, the Nicolsons were receiving advice from Powys, who was at the time Secretary of the Society for the Protection of Ancient Buildings. Powys was largely responsible for work at the Priest's House as well as for the conversion of the old stables into the Long Library. But his greatest contribution to Sissinghurst was the west wall of the Rondel Rose Garden (Figure 16–4). Powys's design for the wall, named "Powys's Wall" by Vita, was inspired by the innovative green palisade in the south garden of Villa Gamberaia in Settignano, Italy, a favorite of the Nicolsons.

By the outbreak of World War II in 1939, the main structure of the garden was essentially finished but would soon acquire a new purpose. In an effort to fulfill their civic responsibility, the Nicolsons transformed their ornamental gardens into agriculturally productive fields. Sissinghurst supplied vegetables to the local community whose family members were off at war. At the end of the war, in the spring of 1945, a celebratory restoration of the garden began.

Harold was mostly responsible for planning the garden's layout, which originates from the old brick walls that belonged to the earlier buildings on the site. These walls established the beginning of a sequence of intimate garden rooms linked by openings that provide both visual and physical connections between the spaces. Each garden room was beautifully planted with a unique composition designed by Vita. For Vita and Harold, the garden was not simply a planted landscape surrounding their home, it *was* their home. The Cottage Garden was like an outdoor living room that extended from the South Cottage, where the couple slept. The White Garden was developed with a space for outdoor dining, which was convenient to the kitchen located in the adjacent Priest's House.

Perhaps the most compelling feature in the garden is the tall brick tower that captivates most visitors as it did Vita when she first visited the Sissinghurst estate (Figure 16–5). From atop the tower, which stands in the

FIGURE 16–5 Vita's Tower, seen from the White Garden.

center of the garden, one can appreciate the organization of the designed landscape against a backdrop of the natural English countryside. Sissinghurst's uniqueness as a garden exists in the way that both formality and informality are brought together in design, producing a wonderfully rich and varied planting scheme organized along a classically formal garden structure.

In 1946, Vita began writing her weekly column for the *Observer.* It was this column that documented her garden's progression and furthered its popularity among garden enthusiasts. As Vita continued to develop and write about her garden, she revealed her romantic personality. Her passion for plants was made evident, as was her desire for the peace and solitude that a garden could bring. Harold's architectural interests and his inclination toward classicism are revealed in his particular contributions to the garden. The Lime Walk, for example, with its classically inspired forms displays his fondness for Italian-stlye gardens. Together this special collaboration between husband and wife proved extraordinarily successful. Vita, although often credited for most of the garden's planning, admitted that the garden would never have been successful without her husband's natural skill as a designer. Harold's understanding of design fundamentals and his appreciation for scale and proportion enabled him to create a strong garden structure, the canvas upon which Vita would paint a marvelous floral composition, combining interesting forms, subtle textures, and striking colors that constituted this splendid garden portrait. While a much-changed social agenda and economic state made lavish gardens of Sissinghurst's scale essentially extinct after World War II, this design has continued to be popular in even the most modest garden plot.

The first of the garden spaces to be experienced by visitors is the front or Top Courtyard. The Top Courtyard is essentially the "welcoming hall," with several alternative garden entrances. The grand view of the tower to the east (directly ahead) encourages many visitors to begin their journey by passing beneath it. A narrow gap in a wall to the right, or south, leads to the rose garden. To the north, or left, an opening in the wall leads visitors into the garden area, "Delos," and near it the White Garden. The walls surrounding the courtyard are bordered by a rich collection of herbaceous and woody plants. These materials are some of the oldest of Sissinghurst's plantings, as this area was one of the first to be addressed by the Nicolsons. The purple border is perhaps the most popular attraction in the courtyard. It displays Vita's talent for combining even the most difficult colors to work with, purple and red. It is indeed a fine accomplishment in creating successful color combinations, which were rather uncommon in gardens in the early twentieth century. Just beyond the tower to the east is a lower courtyard referred to as the "Tower Lawn." The design of this area, a crossing point between two garden axes, was strengthened by the addition of the Yew Walk (c. 1932), which separates the Tower Lawn from the orchard.

To the south is the rose garden, featuring over 85 varieties of roses, many old-fashioned varieties that Vita favored most. This was her most treasured garden, a product of her passion and a true expression of her romantic character. The numerous varieties of roses are planted in formal beds that surround Harold's Rondel located at the center of the garden. The Rondel is laid out as a circle of lawn ringed with yew hedges, broken only by four narrow gaps through which the garden paths travel. Its classic form is successful as an organizing element in the space, and now in its mature state its green walls contribute to the architectural framework of Sissinghurst's collection of garden rooms.

The grove of hazel trees, now known as the "Nuttery," already existed on the site. It seems that the couple's discovery of this collection of trees reaffirmed their affection for the property. The trees recalled a Kent tradition. The hazels were valued for their nuts as well as for their branches, which were used for making wattle fencing. A carpet of many-colored polyanthus, a much-desired flower in Edwardian gardens, once covered the earth beneath the nut trees. The polyanthuses were replaced in the 1970s with woodland plants such as ferns and *anemones* when a nonlimiting soil disease made growing polyanthuses impossible in that area of the garden. The planting in the Nuttery is more natural-looking than anywhere else in the garden. An element of peacefulness and quiet beauty exists in this garden, with a romantic spirit that still captivates visitors.

The Lime Walk, along the southern border of the property, is one of the first areas of the garden to flower in the spring; it was created entirely by Harold. He not only designed this part of the garden but also planted and maintained it throughout most of his life. His goal was to create a classical garden such as those he so admired in Italy. It was designed as a simple and straightforward avenue of pleached limes (lindens) flanked by clipped hornbeam hedges on either side. The path between them was paved with brightly colored concrete slabs, but has since been replaced by natural stone pavers. The full length of the axis is bordered on both sides by a colorful mixture of plantings at their peak in early spring. As Lord points out, the Lime Walk's formality derives from its architectural simplicity with a clear axial focus strengthened by emphasizing the repetition of line with the trunks of the lindens and also by incorporating terra-cotta pots at regular intervals along its length (Lord 1995).

The South Cottage was the very first building the Nicolsons occupied at Sissinghurst. The garden that they planted around it became their own special private space, a garden that they equally tended and adored (Figure 16–6). Inspiration for the selection and placement of plantings came from one of Vita's most admired garden writers and critics, Irishman William Robinson. Champion of the informal garden style, Robinson expressed his ideas for naturalistic planting in several books that he authored, including *The English Flower Garden* (1883) and *The Wild Garden* (1870). These were among Vita's favorite garden making references. Accordingly, many "cottage garden charms" (Brown 1990), as Robinson called them, are found in Vita's and Harold's own garden. The Cottage Garden is a cozy space bursting with color. The oranges, yellows, and reds make the space warm and inviting. Intimacy is achieved not only in the composition of warm colors but also by the presence of hedges and small, ornamental trees lending the area a sense of enclosure and privacy.

FIGURE 16–6 Inspiration for the South Cottage Garden came from one of Vita's most admired garden writers, William Robinson, author of hair The English Flower Garden *and*The Wild Garden.

From the Cottage Garden, the view to the east follows the Moat Walk to the statue of Dionysus. The Moat Walk defines the path, a likely remnant of the medieval manor house that once occupied the site. It is the primary feature of this garden. Its presence as a historic element in the garden led to Vita's and Harold's decision to do little in the way of planting along its length. Opposite the wall, the garden path is bound by a collection of yellow azaleas that both capture one's view and disguise a substantial grade change between the adjacent Nuttery and the Moat Walk.

The landing providing the transition between the Cottage Garden and the steps that descend to the Moat Walk was referred to as the "Crescent." The Crescent succeeds in creating an organized connection between adjacent garden areas, linking the entrances to the Cottage Garden, the Nuttery, and the Lime Walk, while at the same time staging an opportunity for a commanding view down the Moat Walk's axis from a garden bench (a copy of Lutyens's seat) positioned within the semicircular space.

The orchard site was once the location of an old medieval manor house and later a grand Tudor mansion. The old foundations buried in the fields eliminated most ideas for creating a garden. The apple and pear trees growing in the orchard were old and unproductive, but Vita chose to spare them, having a vision for their use. She planted roses aside the trees, using their canopies as natural arbors. Beneath them, a meadow was created, blanketing the ground with yellow daffodils and narcissus in the springtime and wildflowers and grasses throughout the summer months. It is a charming area unlike any other in the garden, a splendid foreground to the distant view of the English countryside.

The White Garden is the most renowned of Sissinghurst's many garden rooms. It has been celebrated as a pioneering accomplishment in the realm of garden design, and indeed it was. Here plants, mostly all-white flowering, are appreciated for their unique forms and textures. Color takes on a subordinate role. The White Garden is situated adjacent to the Priest's House, where Vita's and Harold's children, Nigel and Benjamin, slept. The family also dined there. Dinners were served inside, and lunches were often set up in a corner of the garden beneath a pergola draped with vines.

A small garden plot occupies a distant location to the east of the Nuttery. This rectangular-shaped space enclosed by yew hedges contains over 100 varieties of herbs, a collection begun by Vita in 1938. Vita had an interest in herbs not for their culinary value but for their historical and romantic associations. The spectacular legends associated with these plants intrigued her. The herb garden was separated from the other garden plots and thus tended to function as a more private space for the couple, but especially for Vita.

Vita died at Sissinghurst in the summer of 1962. She left the estate to her younger son, Nigel, with life rights to the South Cottage bequeathed to her husband. Harold passed away in 1968. A year before their father's death, Nigel and his brother, Benjamin, who had inherited other family property, discussed the transfer of the Sissinghurst property to the National Trust. This decision would be made in order to avoid payment of the estate or death duty tax.

FIGURE 16–7 The Gazebo (1969), planned by Nigel Nicolson and architect Francis Pym, is dedicated to the memory of Harold Nicolson. Its conical roof resembles the shape of the Oast houses constructed across the countryside in Kent.

Today the garden is perhaps the most visited public garden in all of England, with over 200,000 visitors each year. It is meticulously maintained by a head gardener and a gardening staff of six to eight full-time people, all with horticultural qualifications and extensive gardening experience.

At Sissinghurst, Vita and Harold preserved the English tradition of a peaceful, agrarian lifestyle (Figure 16–7). The beauty of life in the English countryside is romantically expressed in their own version of an English country garden. Sissinghurst, was never intended by either Vita or Harold as a garden for posterity (Lord 1995), but it has been the inspiration for many generations of cottage gardeners. Though its creation represents a time in history long since passed, its style offers a timeless beauty.

Chapter 17

Landscape Expression: Modern Variations

The English Walled Garden at Chicago Botanic Garden

1. Checkerboard Parterres
2. Pergola Garden
3. Cottage Garden
4. Courtyard Garden
5. Stone Urn
6. Formal Garden
7. Stone Balustrade
8. Vista or Sunken Garden w/Octagon Pond
9. Garden Pavilions
10. Perennial Borders

The English Walled Garden at Chicago Botanic Garden

GLENCOE, ILLINOIS

Many modern gardens borrow from historic tradition, but few are designed to actually reveal the evolution of a particular gardening style. The English Walled Garden at Chicago Botanic Garden, designed in the early 1990s by English garden designer John Brookes, does both.

Brookes's overall design for the garden recalls the seventeenth- and early-twentieth-century tradition of creating separate garden compartments, each with a different theme or aesthetic. The garden is organized as a series of garden rooms that represents various features and design techniques consistent with the English gardening style throughout its evolution. The general effect is a garden of splendid variety and interest. The garden rooms are organized around a framework of walls, hedges, and pathways that provides scale to the spaces and creates a sense of order and coherence in the overall scheme.

The plan is rather formal, with garden spaces laid out according to geometric principles dating back to Renaissance design. Individual sections are designed as rectangular or square plans, joined by garden steps and pathways. There are six individual garden rooms, each representing a different style of the same or a different period. These include the Formal Garden, Vista or Sunken Garden, Checkerboard Garden, Pergola Garden, Cottage Garden, and small urban Courtyard Garden.

The garden's main axis extends from the Formal Garden to the Sunken Garden and then on to the lake, with a secondary axis stretching from the Formal Garden to the round pool between the Checkerboard Garden. The Formal Garden recalls the late-nineteenth- and early-twentieth-century revival of formality that first appeared in English gardens during the seventeenth century. Its design reflects the philosophies popularized by the collaborative work of early-twentieth-century English garden designer Gertrude Jekyll and architect Edwin Lutyens. Their style of gardening combined a strict, formally designed framework with natural-looking, informal drifts of plants.

Various features common to this style of design have been incorporated into Brookes's Formal Garden: a raised terrace bound by a stone balustrade, garden steps, abundant flower borders, statuary, and garden

FIGURE 17–1 An assortment of daisies colors the formal garden on the raised terrace. Courtesy of the Chicago Botanic Garden.

seats (Figure 17–1). The raised terrace, modeled after Italian Renaissance examples, was designed into the English garden during the seventeenth century and then removed during the eighteenth-century boycott of formality. It reappeared in the English garden during the nineteenth century, becoming one of the primary design features. It often was the point from which the entire plan of the garden evolved. These terraces, approached by garden steps flanked on either side by elaborate stone balustrades, established a refined garden area located adjacent to the house.

Brookes's terrace offers a fine prospect of other areas of the garden with views out to the lake and the planted islands beyond. Plantings in the garden are casual. Informal drifts of various colors and textures soften hard garden lines and establish a pleasing contrast to the garden's formal structure. Brookes introduces a medley of bright colors and interesting textures to the garden's borders, mostly with daisies, and also around its central feature, a stone urn, which is yet another traditional element (Figure 17–2). In the revival of the formal garden style during the late nineteenth century, interest in

FIGURE 17–2 A stone urn provides a focal point at the center of the formal garden. Courtesy of the Chicago Botanic Garden.

garden ornaments was revived. Stone urns, vases, and obelisks were integrated into garden arrangements as focal points, or combined with balustrades and steps. We see this same interest in garden ornament in Brookes's formal composition, with the stone urn chosen as the garden's central feature, and elsewhere throughout the various gardens, with vases and stone statues incorporated into his design.

From the Formal Garden, steps descend to the Vista Garden. Beginning in the early 1700s, the dawn of a new era in garden making, designers started to incorporate distant views of the countryside into their garden schemes.

FIGURE 17–3 The long view from the raised terrace across the lake to Evening Island. Courtesy of the Chicago Botanic Garden.

FIGURE 17–4 At the center of the Sunken or Vista Garden is an octagonal pond modeled after a similar one designed by Nathanial Lloyd at Great Dixter. Courtesy of the Chicago Botanic Garden.

FIGURE 17–5 The Octagon Pond at Great Dixter Garden in East Sussex, England.

The Vista Garden recalls this eighteenth-century tradition, with its long views of the natural-looking landscape garden on the opposite side of the lake known as "Evening Island" (Figure 17–3). Brookes designed the Vista Garden as a sunken garden, a purely English invention dating back to the seventeenth century. The central feature of the garden is an octagonal pool surrounded by a stone wall and steps (Figure 17–4). The inspiration for this seems to have come from the sunken garden and octagonal pool at Great Dixter in East Sussex, designed in 1923 by Nathanial Lloyd, for it is quite similar to his plan (Figure 17–5).

Brookes's Sunken Garden is bound on two sides by tall brick walls, each ornamented by a water-spouting gargoyle cast into its façade. Surrounding the terrace is an abundant border of mixed plantings that softens the garden's boundaries. A stone balustrade foregrounds the distant view over the lake. At the far corners of the garden are handsome, classic-looking garden pavilions that Brookes refers to as "Pepper Pots." One of the structures is actually a maintenance building, but the other one offers visitors shelter from the sun and inclement weather (Figure 17–6). Garden pavilions have played a key role in English gardens since the seventeenth century. Their placement on the corners of a garden site was a way to identify the area belonging within the spatial boundaries of a residence. Inside the open pavilion, situated on the southwest corner of the Vista Garden, a pair of garden seats offer a peaceful resting place.

FIGURE 17–6 A classical looking garden pavilion at the corner of the Vista Garden provides shelter and offers a beautiful view across the lake. Courtesy of the Chicago Botanic Garden.

FIGURE 17–7 The Checkerboard Garden, reminiscent of an Italian parterre, *uses plants as geometric forms to create an interesting design feature. Courtesy of the Chicago Botanic Garden.*

On the east side of the garden is what Brookes calls the "Checkerboard Garden," a modern-looking adaptation of the classic *parterre.* Blocks of green boxwood and silver artemisia are combined to form a checkerboard pattern on either side of a circular pool (Figure 17–7). The pool is surrounded by a radiating pattern of cobblestone pavement. Natural materials such as stone and brick are common to English gardens and appropriate to their "natural" style of gardening. Surrounding the pool is a semicircular garden seat, backed by a brick wall. The wall offers a background as well as a surface upon which climbing vines and **espaliers** (trees, often fruiting varieties that are pruned and trained to grow against a wall or other supporting structure) are featured.

Tucked into a quiet corner of the garden, behind the Checkerboard Garden, is a large pergola taken over by wisteria (Figure 17–8). The idea of a pergola in the garden was quite popular in the sixteenth and seventeenth centuries but absent from eighteenth-century naturalized landscapes. Its return to the garden in the late 1800s coincided with a growing desire for more spatial structure in the garden. The pergola, which dates back to Italian Renaissance design, organizes movement through space. It can be a rather peaceful place to linger in the garden. Shaded by a host of colorful and fragrant flowers borne on a tangle of vines, the space is a garden itself with most of its compostion occurring overhead (Figure 17–9). Beneath the trellis of Brookes's Pergola Garden is a garden seat in the style of the popular twentieth-century "Lutyens Bench," designed by Edwin Lutyens.

FIGURE 17–8 The Pergola Garden. Courtesy of the Chicago Botanic Garden.

FIGURE 17–9 View from the pergola to the adjacent Checkerboard Garden. Courtesy of the Chicago Botanic Garden.

The Cottage Garden, one of the most popular garden spaces in Brookes's composition, not only represents a historic garden style, it is a significant part of English culture and lifestyle. Cottage gardening represented the simple, "back-to-nature" lifestyle of nineteenth-century rural families that planted gardens as a means of subsistence. Their gardens were small and typically enclosed by fences, walls, or hedges, intended to keep animals out of the garden. They were planted with a combination of fruits, vegetables, herbs, and ornamental plants. Working families had little time to spend in their gardens, thus they often appeared neglected, sometimes looking rather weedy and unkempt. It was this neglect, the random effect of untamed nature, that

FIGURE 17–10 A grove of apple trees in the Cottage Garden. Courtesy of the Chicago Botanic Garden.

appealed to late-nineteenth- and early-twentieth-century gardeners who reacted against the contrivances of nineteenth-century Victorian exoticism.

The new "English cottage style" evolved as more of an offshoot of the practical rural garden, borrowing from its seemingly wild and random character to create its own identity, one that Brookes refers to as the English gardening interpretation of the term "shabby chic." Vegetables, fruits, and flowers are planted in a beautiful, burgeoning sort of way, all contained within the structure of a formal outline or enclosure (Brookes 2003). Jekyll and Lutyens popularized this style during the late nineteenth and early twentieth centuries. Their gardens inspired many subsequent designs developed in the "cottage way." For instance, several of the most widely known gardens in England, such as Sissinghurst in Kent and Great Dixter in East Sussex, were influenced by the techniques of Jekyll and Lutyens.

Brookes's Cottage Garden is filled with plants spilling out into the brick pathways that wind through the space. A small orchard planting of elegantly knurled apple trees provides structure to the space and recalls the practical aspect of this type of garden (Figure 17–10), as does the interesting collection of terra-cotta and stone pots filled with aromatic herbs (Figure 17–11). These pots, set in various places in the garden, fill the entire space with a pleasant, herbal fragrance. Plants are held together by the garden's vaguely formal structure, mostly derived from the organization of paths. The plants are laid out in drifts rather than singularly, and they are arranged in a complementary fashion. Colors, forms, and textures appear to combine randomly, though in truth considerable planning is involved in this technique of garden making (Figure 17–12).

At the west end of the path that defines the northern boundary of the garden is a small re-creation of a town-like garden shaded by large trees and a rustic, vine-covered arbor (Figure 17–13). During the nineteenth century, cities grew as a result of England's industrial growth. Many families relocated

FIGURE 17–11 Fragrant herbs are grown in terra-cotta and stone pots in the Cottage Garden. Courtesy of the Chicago Botanic Garden.

FIGURE 17–12 The English Cottage Garden is filled with exciting colors and textures. Courtesy of the Chicago Botanic Garden.

FIGURE 17–13 The Town or Courtyard Garden is a quiet, shady retreat. Courtesy of the Chicago Botanic Garden.

FIGURE 17–14 A bold and brightly colored herbaceous border flanks the garden's main pathway. Courtesy of the Chicago Botanic Garden.

to be closer to the factories that employed them. On very small urban properties, they created courtyard gardens that provided a pleasant retreat to nature in what was typically a crowded, sterile living environment. Brookes introduces many shade-tolerant plants into his Courtyard Garden to create a composition that could survive in a typical urban setting, where tall buildings shade everything around them. Many of the plantings are displayed in containers that are decorative and at the same time versatile. Opportunity exists to create variety in this limited space simply by rearranging pots and changing the different types of plants in the various containers. Central to this space is a large lead cistern, dating from 1716, which provides an ideal focal point.

A long, vibrant perennial border dotted with interesting topiaries flanks the main walkway along the garden's west wall. It is a spectacle of color and bold textures. This technique for creating herbaceous borders, developed a century ago by Jekyll, is every bit as popular in English gardens today. Its bold, romantic appearance appeals to many American gardeners, who have since adopted the technique (Figure 17–14).

The English Walled Garden is just a small glimpse of England's long and passionate heritage of gardening. Yet as such, it captures the spirit of the English gardening tradition by presenting a warm, pleasant, and inviting atmosphere in which to take in the many benefits of nature. Its success is due to Brookes's skillful blending of significant elements of the past in a complementary, unified way. It is a garden owed to tradition but valuable to the future as a lesson in the possibilities of creating new gardens inspired by historic achievements.

Chapter 18

Landscape Expression: The Royal Parks of England

London is famous for its historic parks, many of which are owned by the monarchy but completely open to the public. Most of the royal parks were originally deer parks, created as hunting reserves for kings and their nobles. The deer park became popular during the Tudor period, especially with King Henry VIII, who was particularly fond of hunting. However, the deer park had a social purpose as well, providing the elite classes with a place for recreation and entertainment as well as sport. Woods were planted for their beauty as well as their function as cover for game.

Hunting in the deer parks was less popular with the Stuarts, beginning with James I in 1603. However, the deer parks continued to evolve as status symbols, providing a setting for the popular "country lifestyle" of the well-to-do and royal classes. Under James, the public, specifically his courtiers, were permitted limited access into the parks. By the time Charles II became king, most parks were opened to the general public, as they remain today. Two of London's oldest royal parks are Hyde Park and St. James's Park.

Hyde Park
London, England

Hyde Park is one of London's most popular and beautiful public parks. Its 350 acres have survived as an open, green landscape, surrounded by a bustling city. It was a royal hunting park created by King Henry VIII in 1536, when he acquired the property through a trade with the abbots of Westminster. During the next century, its character would change. Hyde Park became a fashionable venue for royal courtiers under the reign of subsequent monarchs and was eventually opened to the general public. The park's grounds, especially the **ring** (a circular drive created by a large enclosure), were frequented by many wealthy visitors. Carriages increasingly circled the course as more and more people sought to display their finery among the wealthiest classes.

From 1665, the year the Great Plague began, the park's social composition would change again. The public gathered on its open lawns, seeking a healthier alternative to the city, where disease was ravaging the population. The plague lasted approximately 18 months, killing more than 100,000 people. This occurrence initiated a period of steady decline in the park. Highwaymen and bandits traveled the park's boundaries, making it a potentially dangerous place.

When William III and Mary II came to the throne toward the end of the seventeenth century, they made an effort to make Hyde Park a safer place, especially since their court was established at the nearby Kensington Palace. The carriage road between the palace and the park was lit with 300 oil lamps, creating the first lighted highway in the country (Figure 18–1). The road was

Figure 18–1 Rotten Row, the first artificially lit "highway" in the country.

Hyde Park Time Line

1536	Henry VIII acquires Hyde Park.
1637	Charles I first opens the park to the public.
1665	Citizens of London flee the city during the Great Plague and set up camp at the park.
1690s	King William III has the carriage road between Kensington Palace and the park lighted, creating the first artificially lighted highway in the country. It was named Route du Roi, or "Rotten Row."
1730s	George II's wife, Queen Caroline, has the park relandscaped and creates the large, winding lake known as the "Serpentine."
1851	The Great Exhibition is held in the park. Sir Joseph Paxton constructs an iron and glass hall in the park, the "Crystal Palace," created especially for the event.
1855	The tradition of public assemblies and demonstrations in the park begins.
1872	The right of assembly is established, making public assemblies legal, and an area called "Speakers' Corner" is set aside for that purpose.
1977	A silver jubilee exhibition is held in the park to honor Queen Elizabeth II's 25 years as queen.
2004	The Princess Diana Memorial Fountain designed by landscape architect Kathryn Gustafson is completed.

Hyde Park

1. Lancaster Gate
2. The Long Water
3. West Carriage Drive
4. North Ride
5. The Meadow
6. North Carriage Drive
7. Bayswater Road
8. Marble Arch
9. Speakers' Corner
10. Dorchester Ride
11. Park Lane
12. Bandstand
13. Wellington Museum
14. Hyde Park Corner
15. Rotten Row
16. The Serpentine
17. Serpentine Road
18. South Carriage Drive
19. West Carriage Drive
20. Albert Memorial

FIGURE 18–2 The sinuous lake called the "Serpentine," created during the eighteenth century for Queen Caroline (1683–1737).

rather unsafe to travel when it first opened in the early 1700s, due to its soft, rutty surface. Its condition resulted in several carriage accidents, one involving King George II's four daughters. By the end of the century, the road was regraded, and a new surface of fine gravel was laid to soften the fall of those cast from their carriages upon it. This new soft surface may account for the derivation of its name, "Rotten Row," from Route du Roi. ("Rotten" is a common Anglo-Saxon word that refers to a soft material.)

In the 1730s, Queen Caroline, wife of George II, took private possession of nearly two-thirds of the park to form her gardens at Kensington Palace. She made improvements to the rest of the grounds by planting trees and linking the existing ponds to form a 35-acre meandering lake, the "Serpentine" (Figure 18–2). The upper reach of this new feature was named the "Long Water" and existed within the boundaries of the palace's gardens. The park became regentrified and grew in popularity as a result of the many public attractions held there. The most popular were equestrian events, various forms of water entertainment on the lake, and a seemingly uncivilized attraction, public executions by way of hangings, which lasted until the late 1700s.

The long-standing tradition of a royal hunt ended toward the end of the eighteenth century, as the park became more of a public venue, with national celebrations held there. The park's popularity continued into the nineteenth century, but extensive use had taken its toll on the grounds. At the turn of the century, Hyde Park was in dire need of improvements. These came in 1823, through the efforts of the Commissioners of Woods and Forests. The commission introduced several physical upgrades as well as fundamental changes in the way the park was managed. Hours of operation were restricted, and the park was now regularly patrolled, making it safer for visitors.

Toward the middle of the century, during the reign of Queen Victoria, the Great Exhibition, a celebration of England's period of industrial and technological greatness, was held in the park. The much-admired Crystal Palace, designed by Sir Joseph Paxton, was specially built in the park to house many of the exhibition's displays.

Public assemblies and demonstrations also became frequent in the park, attracting enormous crowds. As this activity continued, it became apparent that some level of organization was needed. Several attempts were made by local authorities to dissuade public assemblies, for fear of the unruly behavior that the large gatherings sometimes initiated. This was strongly protested and, finally, in 1872, the right of assembly was established, and an area of the park, "Speakers' Corner," was set aside as a public-speaking venue.

St. James's Park
London, England

St. James's Park is the oldest of the eight royal parks in London. It is surrounded by three palaces: Westminster, now the home of the Houses of Parliament; St. James, once the official court; and Buckingham Palace, the home of the monarch since 1837. The park, formerly a marshland, took its name from the nearby St. James's Leper Hospital. Henry VIII, impatient to remove the hospital from view of his Westminster Palace, acquired the site in 1532 and created a deer park. Accordingly, St. James's Palace was named for the hospital upon whose foundation it was built.

James I (1566–1625) made improvements on the land, controlling the floods from the Tyburn stream that flowed through the park. He introduced into the site many exotic animals, such as antelopes, crocodiles, camels, pelicans, and an elephant, many of these acquired as gifts from foreign nobles. Charles I (1600–1649) made some minor changes to the site, but it was his son, Charles II (1630–1685), who dramatically altered the park's appearance.

Shortly after Oliver Cromwell's death (Cromwell ruled England as a republic from 1653 to 1658 after the execution of Charles I, who was sentenced to death by the Parliamentarians in 1649), Charles II returned from his exile in France and was restored to the throne. The king had grown fond of the French formal style and wished to create a similar layout in the park. French landscape designer André Mollet (d. 1665) was hired to redesign the park. A canal, approximately one-half mile long, was built to create an axis, alongside a new avenue called the "Mall." Both were lined with crisp rows of trees and flanked by green lawns. The Mall runs parallel to the Pall Mall, pronounced *"Pell Mell,"* which was a space designated for the game pall-mall. Pall-mall is the English equivalent of *paille maille* (similar to current-day croquet), which the king learned to play while in France. Toward Whitehall another French-inspired technique was employed, the *patte d'oie,* or goosefoot style, layout of avenues. When the renovation was complete, the king opened the park to the public. He continued to frequent the park, where he especially enjoyed watching the waterfowl.

At the dawn of the eighteenth century, the park had a rather shameful reputation. Theft, prostitution, and other crimes were becoming more of a problem, and the park was becoming less of a fashionable venue. In the second half of the century, George III (1738–1820) bought Buckingham House, which had been built by the Duke of Buckingham on the west end of the mall. George IV (1762–1830), upon his accession to the throne in 1820, had Buckingham House upgraded to a royal palace. Architect John Nash (1752–1835) created the palace for the king during the 1820s and also redesigned (and as a result improved) St. James's Park. Nash developed an in-

St. James's Park Time Line

1532	Henry VIII acquires St. James and creates a royal deer park.
1570s	James I establishes a menagerie at the park, with exotic animals from all over the world.
1660s	Charles II has the park redesigned in the French style. The canal is built, avenues of trees are planted, and the Pall Mall is created alongside the park. After the renovation, the park opens to the public.
1761	George III acquires Buckingham House.
1820s	George IV upgrades Buckingham House to a royal palace. Architect John Nash redesigns the park in the popular "English style," removing all traces of formality.
1837	The Ornithological Society of London donates birds to the park and builds a cottage for a bird keeper.
1857	An iron suspension bridge by Matthew Digby Wyatt is built across the lake, replacing Nash's original structure.
1911	The Queen Victoria Memorial by Thomas Brock is erected in front of Buckingham Palace.
1957	Wyatt's bridge is replaced by a new bridge designed by Eric Bedford.
1970	The Cake House, designed by Eric Bedford, is built to house a self-service cafeteria.

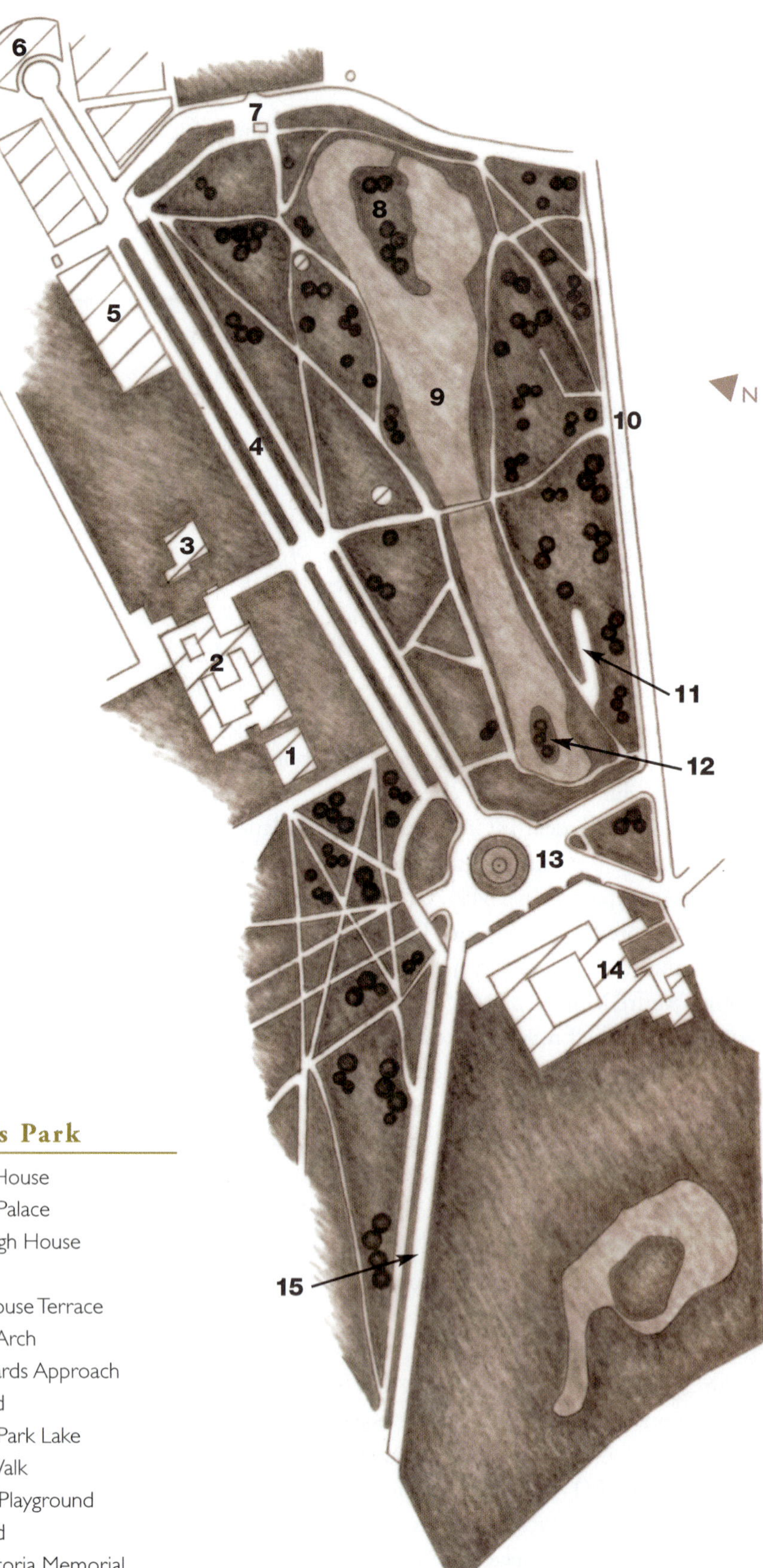

St. James's Park

1. Lancaster House
2. St. James's Palace
3. Marlborough House
4. The Mall
5. Carlton House Terrace
6. Admiralty Arch
7. Horse Guards Approach
8. Duck Island
9. St. James's Park Lake
10. Birdcage Walk
11. Children's Playground
12. West Island
13. Queen Victoria Memorial
14. Buckingham Palace
15. Constitution Hill

formal plan for the park, removing all formality in the then-popular "romantic" style. The formal canal was replaced by a more natural-looking lake, known as the "Ornamental Water," which was surrounded by winding paths and informal plantings of flowering shrubs and shade trees (Figures 18–3, 18–4). The park reflects this style today.

FIGURE 18–3 Nash's natural-looking lake, known as the "Ornamental Water," was part of a nineteenth-century restoration.

FIGURE 18–4 The park's informal style is defined by winding paths and casual groupings of shade and ornamental trees.

Selected Bibliography

The following titles were valuable resources in the preparation of this book and are highly recommended for further reading.

Ackerman, James S. *The Villa: Form and Ideology of Country Homes.* Princeton: Princeton University Press, 1990.

Alberti, Leon Battista. *L'architettura (De re aedificaturia).* Ed. P. Portoghesi, translated by: G. Orlandi, 2 vols. Milan, 1966 (completed in 1452, first published in 1485).

Babelon, Jean-Pierre. *Chantilly.* Paris: Éditions Scala, 1997.

Bajard, Sophie, and Raffaello Bencini. *Villas and Gardens of Tuscany.* Paris: Terrail Editions, 1993.

Batey, Mavis. *Pope, Alexander, The Poet and the Landscape.* London: Barnes Elms, 1999.

Batey, Mavis, and Jan Woudstra. *The Story of the Privy Garden at Hampton Court.* London: Barns Elms, 1995.

Bettini, Giovanni. *Bomarzo.* Terni, Italy: Plurigraf Narni, 1988.

Biddulph Grange Garden. London: Centurion Press, 1999.

Bisgrove, Richard. *The Gardens of Gertrude Jekyll.* Berkeley: University of California Press, 2000.

Blomfield, Reginald. *The Formal Garden in England.* London: Macmillan & Company, 1892.

Bond, James, and Kate Tiller. *Blenheim, Landscape for a Palace.* Phoenix Mill, UK: Sutton, 1997.

Boulding, Anthony. *The History of Hampton Court Palace Gardens.* Surrey: Hampton Court Palace, (date unknown).

Brawne, Michael. *The Getty Center, Richard Meier and Partners.* London: Phaidon Press, 2000.

Braybrooke, Neville. *London Green: The Story of Kensington Gardens, Hyde Park, Green Park, and St. James Park.* London: Victor Gollancz, 1959.

Brookes, John. A Lecture at the Chicago Botanic Garden's Great Gardens Design Symposium, January 17, 2003. Glencoe, Illinois.

Brown, Jane. *Gardens of a Golden Afternoon.* New York: Van Nostrand Reinhold, 1982.

———. *Sissinghurst, Portrait of a Garden.* London: Orion, 1990.

———. *The English Garden through the 20th Century.* Woodbridge, Suffolk, England: Antique Collectors Club, 1999.

Campbell, Malcolm. "Hard Times in Baroque Florence: The Boboli Garden and the Grand Ducal Public Works Administration," in *The Italian Garden,* edited by John Dixon Hunt, pp. 160–201. Cambridge: Cambridge University Press, 1996.

Clarke, George, and Jonathan Marsden, Richard Wheeler, Michael Bevington, and Tim Knox. *Stowe Landscape Gardens.* London: Balding and Mansell for the National Trust, London, 1997.

Coffin, David R. "The Elysian Fields of Rousham." Proceedings of the *American Philosophical Society,* Vol. 130, No. 4, 1986, pp. 406–423.

———. *The Villa D'Este at Tivoli.* Princeton: Princeton University Press, 1960.

———. *The Villa in the Life of Renaissance Rome.* Princeton: Princeton University Press, 1979.

Conniff, Richard. "The House that John Built." *Smithsonian Magazine,* February 2000, Vol. 31, No. 2, pp. 104–114.

Conway, Hazel. *Public Parks.* Buckinghamshire, UK: Shire Publications, 1996.

Covington, Richard. "Renaissance of the Gardens of Versailles." *Smithsonian Magazine,* July 2001, Vol. 32, No. 4, pp. 91–100.

Deal, Joe. *Between Nature and Culture.* Los Angeles: The J. Paul Getty Museum, 1999.

Dernie, David. *The Villa D'Este at Tivoli.* London: Academy Group, 1996.

Devillers, Laurent, Director of Publications. "Fontainebleau: A Royal History." Document édité parle Comité Departemental du Tourisme de Seine-et-Marne. Fontainebleau: France, 2002, pp. 65–66.

Devonshire, Duchess of. *Chatsworth Gardens.* Derby: Derbyshire Countryside, 1999.

———. *Chatsworth.* Derby: Derbyshire Countryside, 1999.

———. *The Garden at Chatsworth.* London: Frances Lincoln, 1999.

Dodd, Dudley, and Kenneth Woodbridge. *Stourhead.* Hampshire, UK: BAS Printers for the National Trust, London, 1990.

———. *Stourhead Landscape Garden.* London: Centurion Press for the National Trust, London, 1981.

Dodd, Dudley, Fred Hunt, and Carola Stuart. *Stourhead Garden.* London: Centurion Press for the National Trust, London, 1985.

Droguet, Vincent. *Fontainebleau: The House of Kings.* Paris: Éditions du Huitième Jour, 2002.

Duggan, Christopher. *A Concise History of Italy.* Cambridge: Cambridge University Press, 1994.

Enge, Torsten, Olaf Enge, and Carl Friedrich Schröer. *Garden Architecture in Europe.* London: 1992.

"England." Microsoft Encarta Online Encyclopedia 2004. <http://encarta.msn.com©1997-2004>.

"England" Microsoft Encarta Reference Library 2003, 1993–2002. Microsoft Corporation.

Eyres, Patrick. "Garden of Apollo and Venus: An Iconographic Speculation." In *New Arcadians Journal 19* (1985), pp. 26–34.
Fitch, James Marston, Joseph Disponzio, Anita Berrizbeitia, Daniel Donovan, Mark Klopfer, and Gary Hilderbrand. *Daniel Urban Kiley, The Early Gardens.* Edited by William S. Saunders, New York: Princeton Architecture Press, 1999.
Fleming, John, Hugh Honour, and Nikolaus Pevsner. *The Penguin Dictionary of Architecture and Landscape Architecture.* New York: Penguin Putnam, 1999.
"France" Microsoft Encarta Reference Library 2003, 1993–2002. Microsoft Corporation. All rights reserved.
Froemke, Susan and Bob Eisenhardt with Albert Maysles. *Concert of Wills, Making the Getty Center* (video). Los Angeles: The J. Paul Getty Trust, 1997.
Galletti, Giorgio. "Un itinerario storico fra i maggiori giardini Medicei." In *Giardini Regali.* Milano, Italia: Electra, 1998, pp. 51–66.
Garnett, Oliver. *Stourhead Landscape Garden.* London: Centurion Press for the National Trust, London, 2000.
Girouard, Mark. *Life in the English Country House.* London: Yale University Press, 1978.
Green, David. *Blenheim Palace.* Norwich, England: Jarrold, 2000.
———. *The Gardens and Parks at Hampton Court and Bushy.* London: W. S. Cowell, 1974.
Gurrieri, Francesco, and Judith Chatfield. *Boboli Gardens.* Florence, Italy: Editrice Edam, 1972.
Gutkind, E.A. *Urban Development in Western Europe; France and Belgium.* New York: The Free Press, A Division of the Macmillan Company, 1970.
Hahn, Hazel. "Jardin des Plantes," in *Chicago Botanic Garden's Encyclopedia of Gardens; History and Design.* Chicago: Fitzroy Dearborn Publishers, 2001, pp. 660–662.
Halliday, F. E. *England: A Concise History.* London: Thames and Hudson, 2000.
Hayden, Peter. *Biddulph Grange.* London: George Philip, 1989.
Hibbert, Christopher. *The House of Medici: Its Rise and Fall.* New York: Morrow Quill Paperbacks, 1980.
———. *The Story of England.* London: Phaidon Press, 2000.
Hilderbrand, Gary R. *The Miller Garden: Icon of Modernism.* Washington, D.C.: Spacemaker Press, 1999.
Hirsch, Jeffrey. *Seeing the Getty Center.* Los Angeles: Public Affairs Department, The J. Paul Getty Trust, 1997.
Hirsch, Jeffrey. *Seeing the Getty Garden.* Los Angeles: The J. Paul Getty Museum, 1998.

Hobhouse, Penelope. *The Story of Gardening.* London: Dorling Kindersley, 2002.

Holmes, Caroline, Editor. *Icons of Garden Design.* London: Prestel, 2001.

Hoog, Simone, and Daniel Meyer. *Versailles Complete Guide.* Paris: Édition Art Lys, 1998.

Hunt, John Dixon. *Garden and Grove: The Italian Renaissance Garden in the English Imagination: 1600–1750.* Philadelphia: University of Pennsylvania Press, 1996.

———. *Gardens and the Picturesque: Studies in the History of Landscape Design.* Cambridge: MIT Press, 1992.

———. *Greater Perfections: The Practice of Garden Theory.* Philadelphia: University of Pennsylvania Press, 2000.

Hunt, John Dixon. "Verbal and Visual Meanings in Garden History: The Case of Rousham," in *Garden History: Issues, Approaches, and Methods,* edited by John Dixon Hunt, Vol. 13. Washington, D.C.: Dumbarton Oaks Research Library and Collection, 1992, pp. 151–181.

Hunt, John Dixon, and Peter Willis, eds. *The Genius of the Place: The English Landscape Garden 1620–1820.* Cambridge: MIT Press, 1997.

Innes-Smith, Robert. *Marlborough.* Derby: English Life Publications, 1992.

"Italy," Microsoft Encarta Reference Library 2003, 1993–2002 Microsoft Corporation. All rights reserved.

Jacques, David. "The History of the Privy Garden," in *The Kings Privy Garden at Hampton Court Palace.* London: Apollo Magazine, 1995.

Jacquin, Emmanuel. *The Tuileries: From the Louvre to the Place de la Concorde.* Paris: Éditions du Patrimoine, 2001.

Jellicoe, Geoffrey, and Susan Jellicoe. *The Landscape of Man.* London: Thames and Hudson, 1995.

Jones, Colin. *France.* New York: Cambridge University Press, 2001.

Kiley, Dan, and Jane Amidon. *The Complete Works of America's Master Landscape Architect.* New York: Bulfinch Press, 1999.

Kostof, Spiro. *A History of Architecture: Settings and Rituals.* New York: Oxford University Press, 1995.

Lablaude, Pierre-André. *The Gardens of Versailles.* London: Zwemmer Publishers, 1995.

Larwood, Jacob. *The Story of the London Parks.* London: Chatto and Windus, 1875.

Lazzaro, Claudia. *The Italian Renaissance Garden.* New Haven: Yale University Press, 1990.

Leszczynski, Nancy. *Planting the Landscape.* New York: John Wiley and Sons, 1999.

Liberman, Alexander. *Campidoglio.* New York: Random House, 1994.

Lord, Tony. *Gardening at Sissinghurst.* London: Frances Lincoln, 1995.

Mader, Gunter, and Laila Neubert-Mader. *The English Formal Garden: Five Centuries of Design.* New York: Rizzoli International, 1997.
Mariage, Thierry. *The World of André Le Nôtre.* Philadelphia: University of Pennsylvania Press, 1999.
Medri, Litta, and Giorgio Galletti. *Boboli Gardens.* Livorno, Italy: Sillabe, 1998.
Meier, Richard, Ada Huxtable, Stephen Rountree, and Harold Williams. *Making Architecture, The Getty Center.* Los Angeles: The J. Paul Getty Trust, 1997.
Mignani, Daniela. *The Medician Villas.* Florence, Italy: Arnaud Ed., 1995.
Miltoun, Francis. *Royal Palaces and Parks of France.* Boston: L. C. Page and Company, 1910.
Mitford, Nancy. *The Sun King.* London: Penguin Books, 1994.
Moggridge, Hal. "Notes on Kent's Garden at Rousham," in *Journal of Garden History,* Vol. 6, No. 3, 1986, pp. 187–226.
Mokhtefi, Elaine. *Paris: An Illustrated History.* New York: Hippocrene Books, 2002.
Montclos, Jean-Marie Pérouse de. *Fontainebleau.* London: Scala Books, 1998.
———. *Vaux le Vicomte.* Paris: Éditions Scala Books, 1997.
Morris, A. E. J. *History of Urban Form.* London: George Godwin, 1979.
Mosser, Monique, and George Teyssot. *The Architecture of Western Gardens.* Cambridge: MIT Press, 1991.
Moughtin, Cliff. *Urban Design: Street and Square.* Oxford: Butterworth-Heinemann, 1995.
Mowl, Timothy. *Gentlemen and Players: Gardeners of the English Landscape.* Phoenix Mill, UK: Stipe, 2000.
Nash, Roy. *Hampton Court: The Palace and the People.* London: Macdonald and Company, 1983.
Newton, Norman T. *Design on the Land.* Cambridge: The Belknap Press of Harvard University Press, 1971.
Nicolson, Nigel, *Sissinghurst Castle: An Illustrated History.* London: Headley Brothers, 1999.
———. *Sissinghurst Castle Garden.* London: Centurion Press, 2002.
Norwich, John Julius. *The Italians: History, Art and the Genius of a People.* New York: Harry N. Abrams, 1983.
Olin, Laurie. *Across the Open Field.* Philadelphia: University of Pennsylvania Press, 2000.
Orsenna, Érik. *André Le Nôtre: Gardener to the Sun King.* New York: George Braziller, 2000.
Pierce, Pat. *London's Royal Parks.* Edited by Barbara Haynes. London: The Royal Parks, 1993.
Pizzoni, Filippo. *The Garden: A History in Landscape and Art.* New York: Rizzoli International Publications, 1999.

Pozzana, Mariachiara. *A Guide to Gamberaia.* Florence, Italy: Edizioni Polistampa, 1999.
Princier, Domaine. *Chantilly.* Paris: Éditions François Bibal, 1989.
Price, Roger. *A Concise History of France.* New York: Cambridge University Press, 2001.
Pugh, Simon. "From Nature as Garden to Garden as Nature," in *New Arcadians Journal, 19* (1985), pp. 8–25.
Quest-Ritson, Charles. *The English Garden: A Social History.* London: Penguin Books, 2001.
Ridgeway, Christopher, and Robert Williams. *Sir John Vanbrugh and Landscape Architecture in Baroque England.* Phoenix Mill, UK: Sutton, 2000.
Robinson, John Martin. *Stowe Landscape Gardens,* The National Trust, 1990.
———. *Temples of Delight, Stowe Landscape Garden.* Hants, England: Pitkin Pictorials in association with the National Trust, London, 1994.
Robinson, William. *The Parks and Gardens of Paris.* London: John Murray, 1883.
———. *The English Flower Garden.* New York: Sagapress, 1984. Originally published 15th edition, London: J. Murray, 1933.
———. *The Wild Garden.* Portland, Oregon: Sagapress/Timber Press, 1994. Originally published 5th edition, London: J. Murray, 1895.
Rogers, Elizabeth Barlow. *Landscape Design: A Cultural and Architectural History.* New York: Harry Abrams, 2001.
Roueché, Berton. "Our Far Flung Correspondents, Pleasant and Living," in *The New Yorker Magazine,* February 1987, pp. 121–125.
Ruggieri, Gianfranco. *Villa Lante.* Florence, Italy: Bonechi Edizioni, 1983.
Samoyault, Jean-Pierre. *Guide to the Museum of the Château de Fontainebleau.* Paris: Éditions de la Réunion, 1994.
Sands, Mollie. *The Gardens at Hampton Court.* London: Evans Brothers, 1950.
Scribner III, Charles. *Bernini.* New York: Harry N. Abrams, 1991.
Sefrioui, Anne. *Vaux le Vicomte.* Paris: Éditions Scala, 1999.
Shepherd, John C., and Geoffrey A. Jellicoe. *Italian Gardens of the Renaissance.* Princeton: Princeton Architectural Press, 1993.
Siciliano, Paul C. Jr. "Fit for a King," in *Chicago Home and Garden Magazine.* Summer 2002. Vol. 8, No. 3. pp. 74–78.
———. "Informal Garden," in *Chicago Botanic Garden's Encyclopedia of Gardens: History and Design.* Chicago: Fitzroy Dearborn, 2001, pp. 631–634.
———. "Italian Pleasures: Villas and Gardens of Renaissance Italy," in *Chicago Home and Garden Magazine.* Spring 2002. Vol. 8, No. 2. pp. 70–75.

———. "Daniel Urban Kiley," in *The Encyclopedia of Gardens: History and Design.* Chicago: Fitzroy Dearborn, 2001, pp.703–706.

Smaus, Robert (garden editor). "A Gardener's Getty," in the *Los Angeles Times.* December 14, 1997, Section K, K1–K2.

Soderstrom, Mary. *Recreating Eden, A Natural History of Botanical Gardens.* Canada: Véhicule Press, 2001.

Spence, Joseph. *Stourhead Landscape Garden.* The National Trust, 2000, p.18.

Steebergen, Clemens, and Wouter Reh. *Architecture and Landscape.* New York: Prestel, 1996.

Sternfeld, Joel. *Campagna Romana:* New York: Alfred A. Knopf, 1992.

Strong, Roy. *The Renaissance Garden in England.* London: Thames and Hudson, 1998.

Sturgis, Matthew. *Hampton Court Palace.* London: Macmillan, 2001.

Switzer, Stephen. *Ichnographia rustica, or, The nobleman, gentlemen, and gardener's recreation.* New York: Gardand, 1982.

Taylor, Patrick. *The Garden Lover's Guide to France.* New York: Princeton Architectural Press, 1988.

Thacker, Christopher. *The History of Gardens.* Berkeley: University of California Press, 1997.

Thurley, Simon. *Hampton Court Palace,* London: Historic Royal Palaces, 2000.

———. "William III's Privy Garden at Hampton Court Palace: Research and Restoration," in *The King's Privy Garden at Hampton Court Palace.* London: Apollo Magazine, 1995.

Trachtenberg, Marvin, and Isabelle Hyman. *Architecture: From Pre-History to Post Modernism.* Upper Saddle River, NJ: Prentice Hall, 1986.

———. *Dominion of the Eye: Urbanism, Art, and Power in Early Modern Florence.* Cambridge: Cambridge University Press, 1997.

Turner, Roger. *Capability Brown and the Eighteenth Century English Landscape.* New York: Rizzoli, 1985.

"United Kingdom." Microsoft Encarta Online Encyclopedia 2004. <http://encarta.msn.com@1997-2004>. Microsoft Corporation. All rights reserved.

Van der Ree, Paul, Gerrit Smienk, and Clemens Steenbergen. *Italian Villas and Gardens.* New York: Prestel, 1993.

Walker, Peter, and Melanie Simo. "The Lone Classicist," in *Invisible Gardens, The Search for Modernism in the American Landscape.* Cambridge, Massachusetts: MIT Press, 1994, pp. 170–197.

Walpole, Horace. "The History of the Modern Taste in Gardening." In *Gentleman and Players: Gardeners of the English Landscape,* edited by Timothy Mowl. Phoenix Mill, UK: Sutton, 2000.

Weaver, Lawrence. *Houses and Gardens by E.L. Lutyens.* Suffolk, UK. Antique Collectors Club, 2001.

Webster, Constance A. "France," in *Chicago Botanic Garden's Encyclopedia of Gardens: History and Design.* Chicago: Fitzroy Dearborn, 2001, pp. 472–478.

Weiss, Allen S. *Mirrors of Infinity.* New York: Princeton Architectural Press, 1995.

Weschler, Lawrence. *Robert Irwin Getty Garden.* Los Angeles: Getty Publications, 2002.

Wharton, Edith. *Italian Villas and Their Gardens.* New York: The Century Company, 1904.

White, Philip. *Hestercombe Garden.* United Kingdom: Hestercombe Gardens Project, 1999.

Whittle, Elisabeth. "Luxembourg Gardens," in *Chicago Botanic Garden's Encyclopedia of Gardens, History, and Design.* Chicago: Fitzroy Dearborn, 2001, pp. 831–833.

Wilhide, Elizabeth. *Sir Edwin Lutyens, Designing in the English Tradition.* London: Pavilion Books, 2000.

Williams, Harold, Bill Lacy, Stephen Rountree, and Richard Meier. *The Getty Center Design Process.* Los Angeles: The J. Paul Getty Trust, 1997.

Wittkower, Rudolph. *Bernini.* London: Phaidon Press, 2000.

Woodbridge, Kenneth. *The Stourhead Landscape.* London: Centurion Press for the National Trust, London, 2001.

Wright, D. R. Edward. "Some Medici Gardens of the Florentine Renaissance: An Essay on Post-Aesthetic Interpretation," in *The Italian Garden,* edited by John Dixon Hunt. Cambridge: Cambridge University Press, 1996.

Young, Geoffrey. *Walking London's Parks and Gardens.* London: New Holland Publishers, 1998.

Zucker, Paul. *Town and Square.* Cambridge: MIT Press, 1973.

Index

C

D

E

F

G

H

I

M

n

o

p

Q

R

S

W

Y